Biotechnology Demystified

Sharon Walker, Ph.D.

McGraw-Hill

New York Chicago San Francisco Lisbon London
Madrid Mexico City Milan New Delhi San Juan
Seoul Singapore Sydney Toronto

The McGraw-Hill Companies

McGraw-Hill books are available at special quantity discounts to use as premiums and sales promotions, or for use in corporate training programs. For more information, please write to the Director of Special Sales, Professional Publishing, McGraw-Hill, Two Penn Plaza, New York, NY 10121-2298. Or contact your local bookstore.

Biotechnology Demystified

4567890 DOC DOC 019

ISBN -13: 978-0-07-144812-3
ISBN -10: 0-07-144812-8

Sponsoring Editor
Judy Bass

Editorial Supervisor
Janet Walden

Project Manager
Ronald D'Souza

Copy Editors
Mark Karmendy and William McManus

Proofreader
Sunita Dogra

Indexer
Jack Lewis

Production Supervisor
Jean Bodeaux

Composition
TechBooks

Illustration
Pam Winslow, Raul Garcia, and Lyssa Wald

Cover Series Design
Margaret Webster-Shapiro

Cover Illustration
Lance Lekander

This book is dedicated to Diane Nicholson and to Red Cloud—they shaped my view of life and love.

ABOUT THE AUTHOR

Sharon Walker, Ph.D, is a Diplomat of the American Board of Toxicology (DABT) and has done extensive research in various areas of biomedicine. Dr. Walker has many years' experience as a safety analyst for high hazard operations, and has taught graduate courses in immunology, epidemiology, cell biology, and statistics.

ACKNOWLEDGMENTS

I gratefully acknowledge the substantial contribution of Bret Wing, who provided a thorough and thoughtful review of this book. I would also like to thank David McMahon who gave invaluable guidance during the development of this book. Finally, many thanks to Lee, Brad, and Lindsey for their encouragment, and to my Uncle Ed and Aunt Sarah for their support.

CONTENTS

INTRODUCTION

Dear Bioengineer to be—I wrote this book for you (and Pam and Raul illustrated this book for you). Personally, as I contemplate these subjects, I experience a sensation similar to the one I had when I watched Neil Armstrong step onto the lunar surface. I would describe it as a new vantage point; not of another point in space, but of our own world. It's another perspective—just like looking back toward Earth from a lunar orbit.

As you move into bioengineering, you may be overcome by a feeling that you have trespassed into forbidden territory just because of the enormous potential for change. Characteristics can and have jumped surrealistically among unrelated species. As you will learn, oil from palm trees can now be produced by rapeseed plants, and pig fetuses can glow like fireflies. We have our hands in the very clay of life. Life is no longer immutable, not only in the characteristics of a species but potentially real time, in the genetics of individuals.

Behind this subject is intense human drama. Some of the drama has reached the popular press, including the emotional debates over embryonic stem cells and the spectacular collapse of the Korean project that claimed to have developed cloned human cell lines. Some of the drama was behind the scenes, such as the race to decipher the human genome—a race between a huge, long-standing government project and an upstart private enterprise that had the audacity to claim the project could be compressed into much shorter time. Some of the drama is yet to come, as we face the potential of designer babies and production of super-athletes. Certainly a basic understanding of this technology and its potential should be a part of the education of all of us, because it is changing our world.

The subject of bioengineering is easier to follow if you have a basic background in science. However, the early chapters summarize the essential information on biological systems such that you should be able follow the descriptions of the technology. The intent was to give you enough of the basic terminology to enable you to understand the literature on this subject, without overwhelming you, dear reader, with foreign words.

This book is about enormous promise and unsettling risks. It involves unique ethical issues and questions about appropriate boundaries for commercial enterprises, if there are any such boundaries. It is about your world now and to come.

CHAPTER 1

Biomolecules and Energy

As a bioengineer, you will be using the systems, subsystems, and molecular units devised by nature towards your own ends. Of course, the more you understand about the natural way of things, the more effective you will be in using and modifying nature. One of the huge benefits reaped from the recent mushrooming of biotechnology initiatives is an unprecedented leap in the knowledge of cellular systems. Before we discuss the recent advances in our knowledge, let's review the basics.

Before you can begin, you must have some knowledge of what you have to work with. First, we need to do a quick review of molecular forces and then we'll describe the building blocks for constructing the most complex of all systems—a living organism.

Review of Molecular Forces

You will need a basic knowledge of how molecules interact to understand how cells function. We don't need to delve into chemistry very far, but you should understand how atoms stick together to form molecules. *Biomolecules* (molecules in biological systems) abide by the same physical laws as any other physical structure.

There are two ways that atoms can associate with one another to form molecules, through *ionic* or *covalent* bonds. To understand this review, you need to know that positively-charged protons exist in the nucleus of an atom, together with neutral neutrons. The negatively-charged electrons travel in shells around the nucleus. Atoms have the same number of electrons as protons, so the atom is neutrally charged. Simple enough.

For the discussion of the electron shells, envision the shells as a series of shelves that must be filled in a certain order. Each shell can accommodate a specific number of electrons. Hydrogen, for example, has one electron in the first shell that exists—the shell with the lowest possible energy state. (Nature seeks the lowest energy state available.) There is one electron because hydrogen has only one proton and the number of electrons in an atom equals the number of protons. The first shell needs to have two electrons before it is full; yes, it wants one more electron than it has protons. So, hydrogen will react easily with other molecules that want to give up an electron. Contrast the behavior of hydrogen with that of helium. Helium has two electrons. The electron shell is perfectly happy, and helium will not react with anything.

When the first shell is full, the next shell, with a higher energy level, receives the additional electrons. The next element is lithium, which has three protons and three electrons. The first shell has two electrons and is stable. However, that third electron in the outer shell wants a buddy badly, so lithium is highly reactive and will give up the third electron at the drop of a hat.

Let's use hydrogen and lithium to explain the difference between ionic and covalent bonds. *Ionic bonds*, shown in Figure 1-1, are formed when one atom gives up an electron to another atom. Lithium will donate its extra third electron to another atom that is short a full house, such as chlorine. When this reaction occurs, lithium assumes a positive charge because the atom has lost the negatively-charged electron, and chlorine receives a negative charge by picking up one more electron that it has positively-charged protons. If this reaction occurs while these atoms are in solution, the positively-charged and the negatively-charge chlorine will float around independently of one another. If you keep adding lithium and chlorine, eventually there are so many atoms that the water cannot keep them apart. The opposite charges will cause these atoms to associate with one another and form lithium chloride salt. The lithium chloride salt is a molecule; however, the forces keeping these two atoms together are not strong and can be easily disrupted by adding more water. And an atom or molecule with a stronger charge can definitely compete for one of the

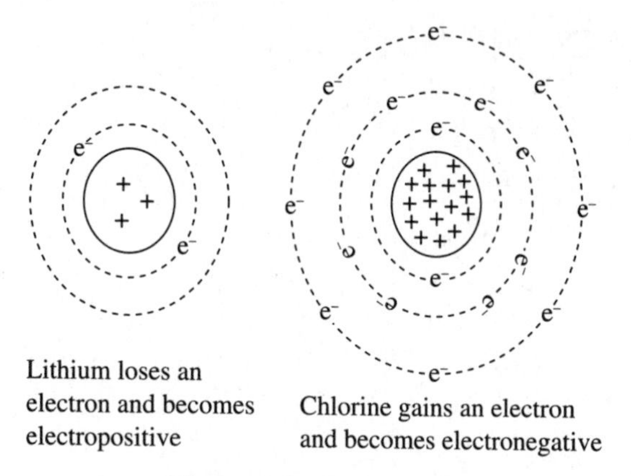

Lithium loses an
electron and becomes
electropositive
Chlorine gains an electron
and becomes electronegative

Figure 1-1 Ionic bonds

"partners." Table salt (sodium chloride) is a familiar example of a molecule that forms ionic bonds in this fashion.

Covalent bonds, shown in Figure 1-2, are formed when atoms share an electron; the electron actually belongs to both molecules. Covalent bonds are much stronger than ionic bonds. Let's return to our example of hydrogen. Hydrogen, as you recall, is one electron short of a full house. Two hydrogen atoms will share their electrons, forming the hydrogen gas, or H2. There are two protons, one per hydrogen atom, and two electrons, shared equally by these atoms.

Often, one atom has a stronger pull than the other, resulting in "polar" regions on the molecule. Consider water. Water is formed when oxygen picks up two electrons, one from each of two hydrogen atoms, and shares these electrons with the hydrogen atoms. However, the sharing is not equal, as shown in Figure 1-3.

Formation of Hydrogen Bonds

Hydrogen bonds are formed between the hydrogen in one molecule and an electronegative molecule in another molecule. Hydrogen tends to form molecules with elements that are strongly electronegative, meaning that the hydrogen bears a partial

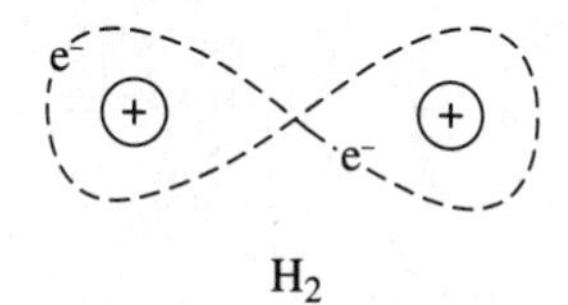

H_2

Figure 1-2 Covalent bonds

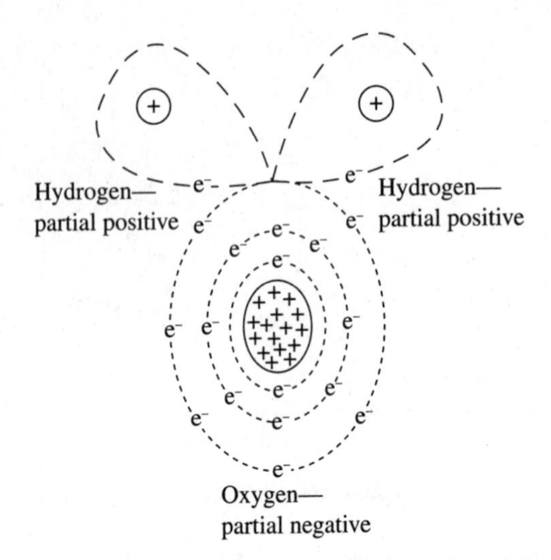

Figure 1-3 Formation of poles in a covalently bonded molecule

positive charge. This positive charge causes the formation of hydrogen bonds. Water is the perfect example of hydrogen bonds.

Consider two water molecules coming close together, as in Figure 1-4.

The hydrogen is strongly attracted to the electronegative poles of the oxygen. Notice that oxygen has two poles, corresponding to the two shared electrons. Hydrogen bonds have about a tenth of the strength of an average covalent bond. They are easily broken and reformed. However, they are significantly stronger than an ordinary dipole-dipole (ionic) interaction. Each water molecule can potentially form four hydrogen bonds with surrounding water molecules. Water

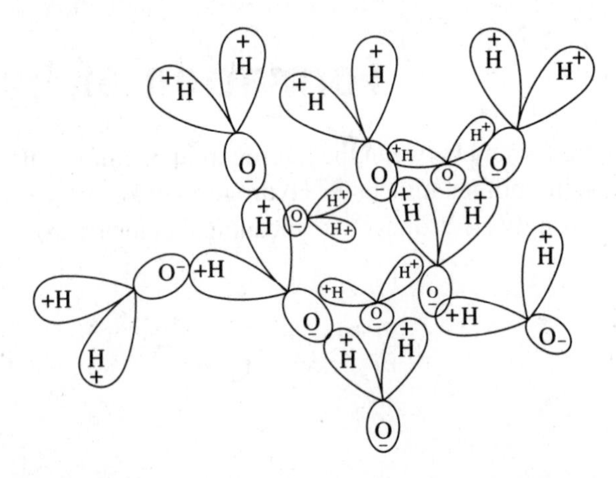

Figure 1-4 Water and the formation of hydrogen bonds

forms a matrix as the dipoles form between the molecules. This is why water molecules appear as though they are trying to cluster together and push out anything in the way.

Biomolecules are formed through covalent bonds, but ionic bonds are important in their interactions and in the relationship between different sections of a given molecule. Hydrogen bonds are very important in the interactions of biomolecules. Many biomolecules literally bristle with hydrogen. Among other things, the formation of hydrogen bonds is a primary force in the way proteins fold in on themselves.

Building Blocks and Reagents: Lipids, Carbohydrates, Proteins, and Nucleic Acids

In general, you have four atoms to work with in forming biomolecules: carbon, oxygen, nitrogen, and hydrogen. (Note: some cellular molecules contain metals, and sulfur is frequently present in disulfide bonds.) These elements form only four types of materials to use in your design of this organism: lipids, carbohydrates, proteins, and nucleic acids. All of these materials build off of a backbone of carbon. Carbon is wonderful because it so readily picks up the other three molecules by forming covalent bonds. Each carbon atom can attach up to four of the other atoms or attach other carbon atoms as well. Carbons also will form a ring where the carbons each share one or two electrons. In the diagrams shown in Figure 1-5, the sharing of two electrons is depicted by a double line, whereas the sharing of only one electron is depicted by a single line.

A biomolecule typically consists of a number of similar units hooked together like a structure made of LEGOs®. The individual building unit is known as a *monomer* and the entire structure is a *polymer*. As we go through this discussion, remember that modular assembly is the norm. Polymers may consist of repeating units of the

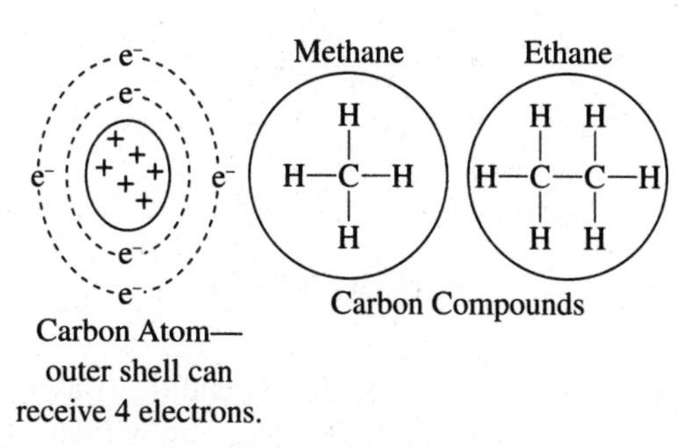

Figure 1-5 The wonderful carbon atom

same material or the units may be unique. Sometimes a polymer will have a given function and knocking off one monomer unit will give it a different function.

There are four classes of biomolecules based on their structure and function—lipids, carbohydrates, proteins, and nucleic acids.

LIPIDS

Individual cells owe their form to the physical behavior of *lipids*, which are the only type of molecule identified only by its physical behavior. Lipids are *hydrophobic*, meaning that they do not dissolve in water (whereas *hydrophilic* molecules are "water loving" and can dissolve in water). This characteristic makes lipids ideal for the formation of membranes. Lipids form large polymers that provide a mechanism for high caloric-content storage, in the form of fats (Figure 1-6 shows an example fat molecule). Some lipids influence biochemical reactions in the cell; in other words, they are bioactive. An example is cholesterol.

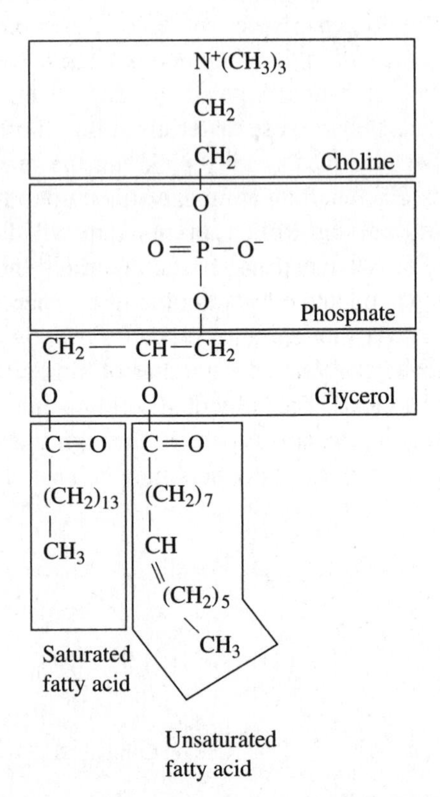

Figure 1-6 Example fat molecule: phosphatidylcholine (Based on Figure 2-7 of *Schaum's Outlines Molecular and Cell Biology* by William D. Stansfield, Jaime S. Colome, and Raul J. Cano, McGraw-Hill, 1996)

Lipids are the primary constituents of the cell membrane. See Chapter 2 for a discussion on how the hydrophobic nature of lipids results in the cell membrane structure.

CARBOHYDRATES

Carbohydrates are the most important source of energy for living organisms and have a formula that is some multiple of CH_2O. For all carbohydrates, if you remove water, only a carbon remains. For multicellular organisms, such as us, the fundamental energy source is a five-carbon sugar known as glucose, a small carbohydrate monomer. If the small monomers are stacked up into large polymers, forming complex carbohydrates (see Figure 1-7 for an example), the resulting molecules can provide structural units. Examples are cellulose and collagen. Other *complex carbohydrates* (composed of large polymers containing many similar monomers hooked together) are used for energy storage. The carbohydrate energy storage molecules are starch in plants and glycogen in animals. This energy storage form has a lower caloric content than fat but is easier to access.

PROTEINS

Lipids and carbohydrates tend to form large, monotonous polymers. *Proteins* become much more interesting. In fact, proteins are so interesting and so diverse that it was difficult to convince researchers in the mid-1900s that proteins were not the information storage molecules for the organism. (We now know that nucleic acids are the information storage molecules.) Proteins form some structures; however, the most important function of proteins is to orchestrate the biochemical reactions that occur in the cell.

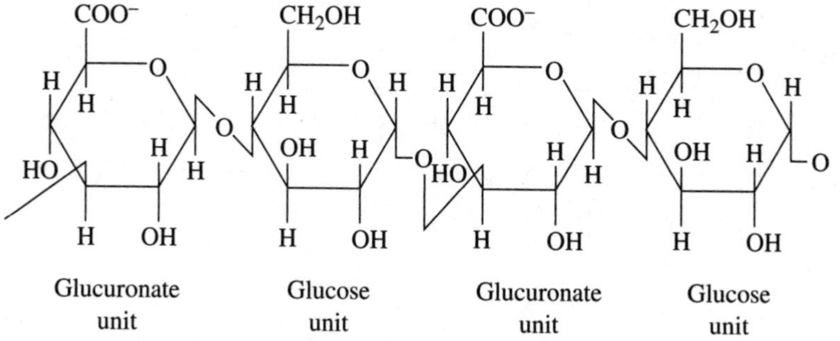

Glucuronate unit — Glucose unit — Glucuronate unit — Glucose unit

Figure 1-7 Example complex carbohydrate (Based on Figure 2-6 of *Schaum's Outlines Molecular and Cell Biology* by William D. Stansfield, Jaime S. Colome, and Raul J. Cano, McGraw-Hill, 1996)

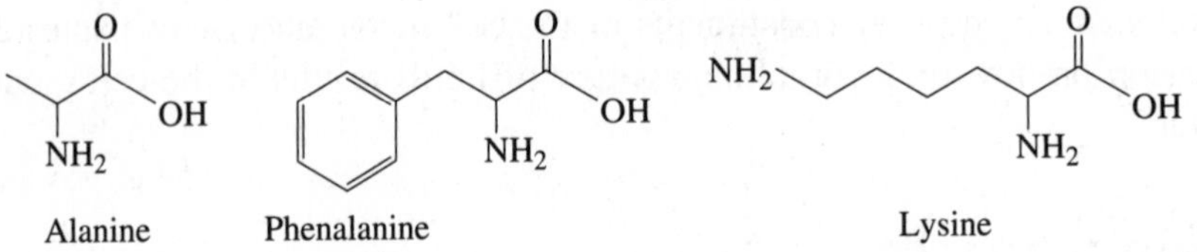

Alanine Phenalanine Lysine

Figure 1-8 Examples of amino acids

Proteins are made of *amino acids*, examples of which are shown in Figure 1-8. These are distinct from carbohydrates and fats because they contain nitrogen. All proteins contain a backbone structure with two carbon atoms and a nitrogen atom. The nitrogen has two hydrogen atoms attached, and the carbon on the end has an oxygen molecule and a hydroxyl group (OH) attached. The middle carbon provides the diversity among proteins because, theoretically, almost anything could be attached. In practice, there are only 20 amino acids, 12 of which the human body can manufacture. The other eight we must consume and are called *essential amino acids*.

Amino acids link together by forming a *peptide bond*, as shown in Figure 1-9. Note that the formation of the bond involves the atoms at the end but leaves that middle carbon atom to retain the characteristic side chains. Protein molecules are large, with the average one containing about 300 amino acids. With 20 amino acids available per position, there are thousands of different proteins that could be formed from 300 amino acids.

The protein is formed as a long chain with numerous side chains, which may be hydrophobic or hydrophilic. The side chains may attract one another by electrical charge or may react chemically. Hydrogen bonds form between side chains. Depending on which side chains are present and in what order, the molecule folds in on itself to form a particular shape. Proteins have a primary structure (linear), and as the molecule begins to fold in on itself, a secondary, tertiary, and a final shape form. This shape is important in allowing the protein to carry out its particular function. Chapter 8 will discuss protein metabolism (the production, use, and destruction of

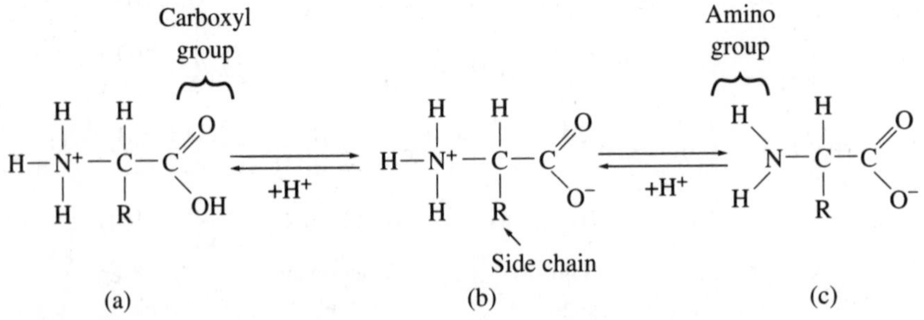

Carboxyl group

Amino group

Side chain

(a) (b) (c)

Figure 1-9 Peptide bond (Based on Figure 2-9 of *Schaum's Outlines Molecular and Cell Biology* by William D. Stansfield, Jaime S. Colome, and Raul J. Cano, McGraw-Hill, 1996)

proteins) in detail because of proteins' importance in cell function and the potential for a bioengineer to intervene in the way proteins are handled by the organism.

NUCLEIC ACIDS

Nucleic acids were discovered in 1868 by Friederich Miescher. The name "nucleic acid" is of historic derivation; nucleic acids are simply acidic molecules that early scientists found in cell nuclei. The discussion of the structure of nucleic acids tends to be confusing because the most important part of the structure is composed of organic bases. There are five of these bases, four found in deoxyribose nucleic acid (DNA) and a fifth found in ribose nucleic acid (RNA). The four bases in DNA are *adenine*, *guanine*, *thymine*, *cytosine*, and the fifth base found in RNA is uracil. Although these names mean little, it is best to familiarize yourself with them because four of them (excluding uracil) are the all-important code of life. Using information encoded in these four compounds, all structures found in living organisms can be constructed. A complete explanation of how this is done is provided in Chapter 3. In that chapter, we will actually break the code—the instruction set for all life forms on earth.

The bases are attached to a sugar. The five-carbon sugar is *ribose*, for RNA. A similar sugar, except that there is one less oxygen, is *deoxyribose*, for DNA. When a base is linked with the sugar, the resulting molecule is known as a *nucleoside*. The sugar picks up a phosphate on the fifth carbon, which makes the whole thing acidic. At this point, the molecule is known as a *nucleotide*. If you can only remember one of these names, remember nucleotide—base and phosphorylated sugar. The sugars link together when an oxygen molecule from the phosphate group on one sugar bonds with a carbon molecule on another sugar. This is called a *phosphodiester linkage*. The second sugar is *phosphorylated* and links with a third sugar, and so on. The linkage of the phosphorylated sugars results in a scaffolding that support the organic bases. The sugar structure forms the railing for the familiar spiral staircase that is DNA. The bases stick to the inside, forming the steps of the staircase.

Let's build it, as shown in Figures 1-10 through 1-13. First, Figure 1-10 shows your sugar. Figure 1-11 adds a phosphate group. Figure 1-12 then links the sugars together, and finally, Figure 1-13 takes a base and attaches it to the sugar.

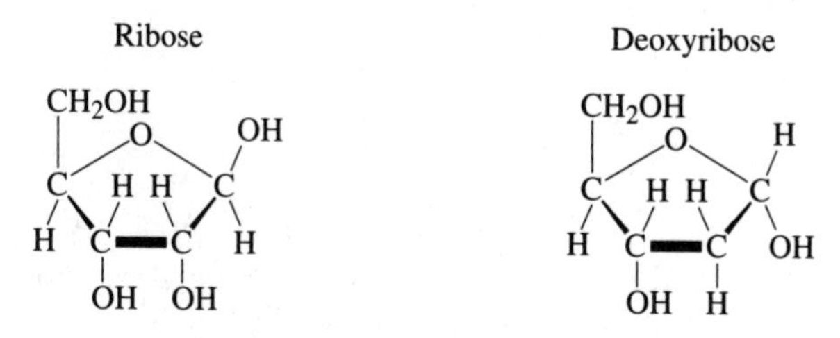

Figure 1-10 Ribose and deoxyribose

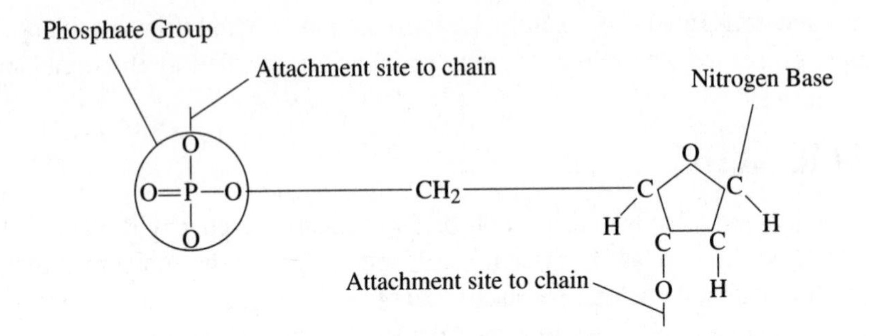

Figure 1-11 Phosphorylated deoxyribose

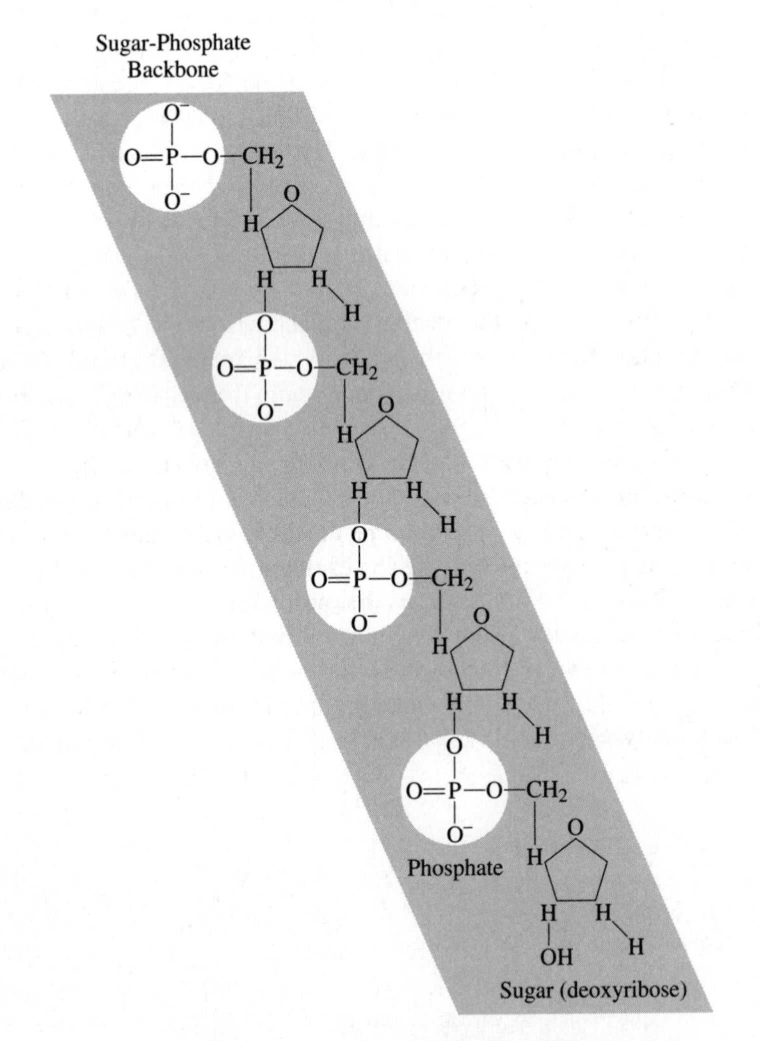

Figure 1-12 The scaffolding for DNA

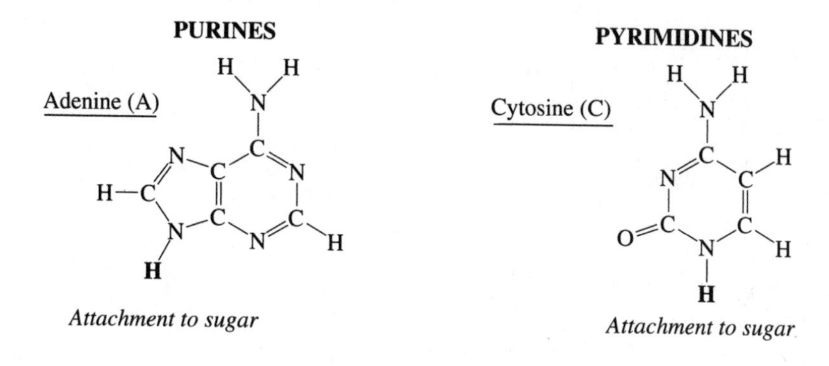

Figure 1-13 The organic bases

Again, as the phosphorylated sugars form the backbone of the molecule, the bases stick to the inside like steps in a staircase, as in Figure 1-14. You need to remember that most DNA is an association of two strands, so there is scaffolding on the right and left and the bases form pair on the inside.

Like most of the phenomena we have discussed, the success of this operation depends on the operatives fitting together physically. Remember that these molecules form unique, complex shapes that just fit the compounds that they react with or bond to. The bases within the DNA molecules are arranged such that the forming of hydrogen bonds is enabled, like linking arms. The shapes of the molecules are such that adenine always links with thymine and guanine links with cytosine, through hydrogen bonding. Even though one hydrogen bond is not particularly powerful, the series of them forms a strong structure, kind of like a zipper. The bonds create a torque that twists the helix on each side, resulting in the familiar *double helix.* Hydrogen bonds are great because the hydrogen bonds are strong enough to hold the structure together, yet weak enough that separating the DNA strands doesn't take a lot of energy. You have to separate the strands when you intend to duplicate the DNA.

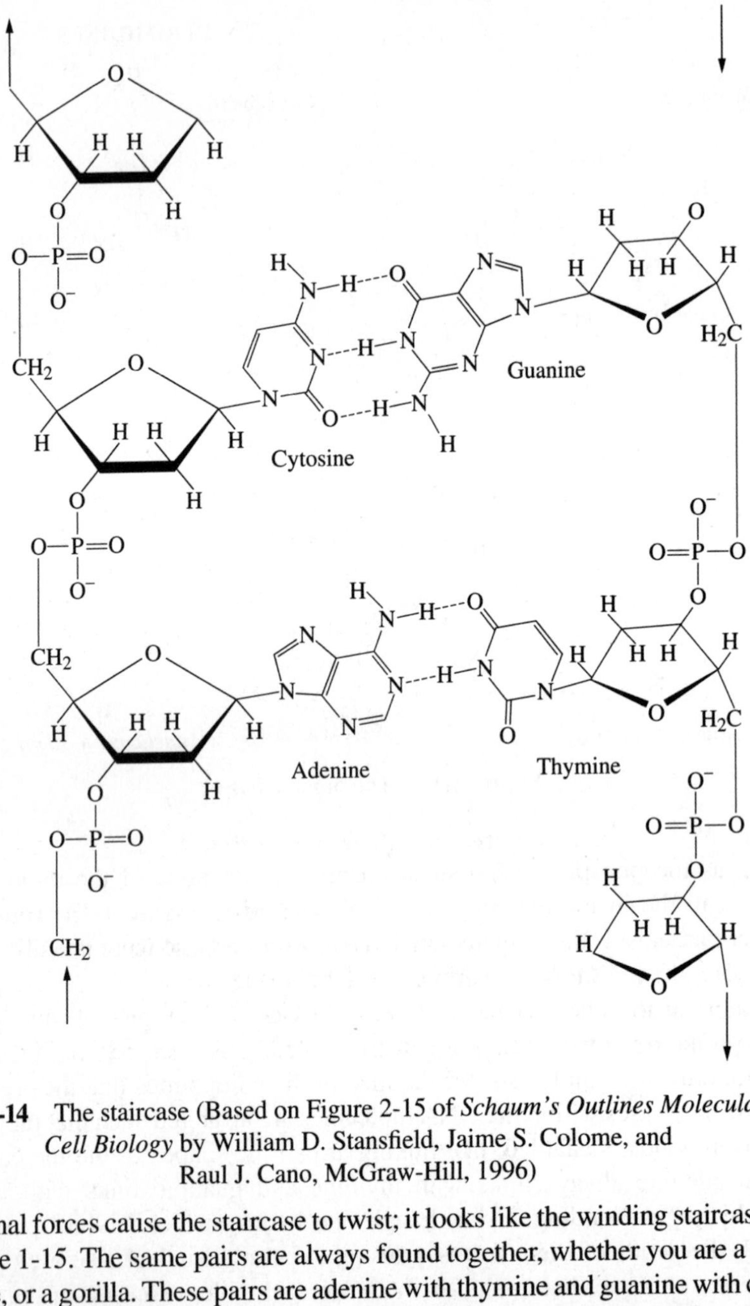

Figure 1-14 The staircase (Based on Figure 2-15 of *Schaum's Outlines Molecular and Cell Biology* by William D. Stansfield, Jaime S. Colome, and Raul J. Cano, McGraw-Hill, 1996)

Internal forces cause the staircase to twist; it looks like the winding staircase shown in Figure 1-15. The same pairs are always found together, whether you are a bacillus, an algae, or a gorilla. These pairs are adenine with thymine and guanine with cytosine. In RNA, the thymine is replaced with uracil. This is a very important rule and is referred to as *complementary base pairing.* Early scientists realized that these compounds were always found in the same amount. Currently, we know how they pair up to build a double strand of DNA from a single strand...but that comes later.

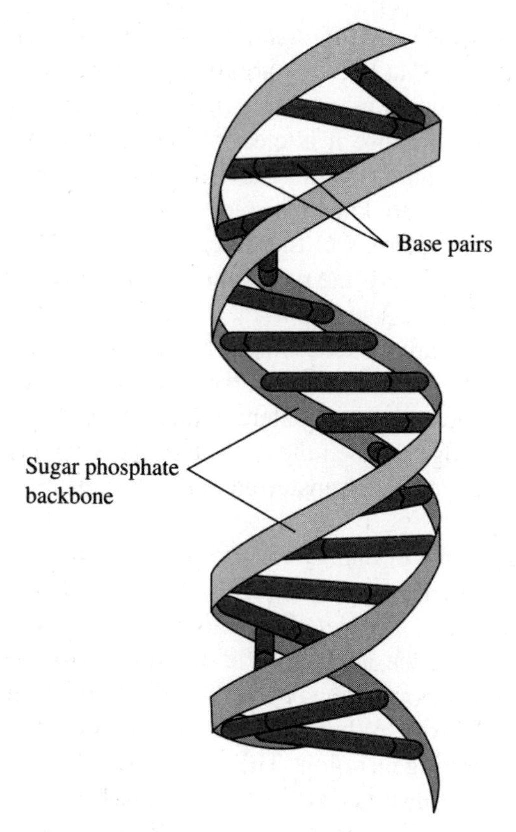

Figure 1-15 The double helix

There are other important nucleotides, the most notable of which is ATP, or adenosine triphosphate. ATP is an energy storage opportunity that we will encounter in a subsequent section.

Now you have the basics to construct a cell.

Energy

The energy for life comes from the exchange of electrons between molecules in a reaction that releases energy. Specifically, the energy for the overwhelming majority of life on earth comes from the *oxidation* of a reduced form of carbon. As luck would have it, this particular reaction releases an unusually large amount of energy. But you know this! You oxidize a reduced form of carbon in your fireplace to create heat! Let me explain.

First, let's review what *oxidation/reduction* means. Oxidation is named after oxygen (duh). Substances that have the ability to oxidize other substances are said to be *oxidative* and are known as *oxidizing agents*, *oxidants*, or *oxidizers*. Oxygen is a very good oxidizer because it is electron-hungry and will remove electrons from another substance. Substances that have the ability to reduce other substances are said to be *reductive* and are known as *reductive agents*, *reductants*, or *reducers*. The reductant transfers electrons to the other substance. This is very difficult to keep straight because ultimately, the reductant will have been oxidized and the oxidant will have been reduced. The important thing to remember for our purposes is that oxygen will oxidize carbonaceous material. The most complete reaction results in the formation of CO_2. In your fireplace, the carbon in the wood is found in molecules containing a large hydrogen content, which means that the carbon is taking electrons from the hydrogen. When the oxygen in the air oxidizes this carbon into carbon dioxide, the electrons are transferred to the oxygen and you benefit from the large energy release.

The most efficient form of energy for life-forms is called *aerobic metabolism*. The specific reactions are captured in a series of events known as the *Kreb's Cycle* or *tricarboxylic acid cycle*. The details of this cycle are beyond the scope of this book. Note, however, that the Kreb's Cycle doesn't use oxygen; that comes later. The Kreb's Cycle generates reducing power by adding hydrogen to small nucleotide molecules (NAD and FAD). When this reducing power meets oxygen, there is potential for large energy generation. The oxidation/reduction reaction involving oxygen occurs in the electron transport chain, so named because electrons are handed off in a series of reactions between molecules. This series of reactions requires oxygen. With each handoff, the energy produced drives the production of adenosine triphosphate (ATP). The phosphate bonds in ATP take significant energy to form and release a lot of energy when they are broken. So, you can see that the cell doesn't burn carbon like you do in your fireplace. Instead, the energy that you would feel as heat from your fireplace is used to create these high-energy phosphate bonds in cells. Think of the convenience! The energy is stored in a highly accessible form (ATP) and can be used whenever needed by the cell. In the end, you have this formula:

$$C_6H_{12}O_6 \text{ (glucose)} + 6O_2 = 6CO_2 + 6H_2O + 36 \text{ ATP}$$

Now, after playing up the role of oxygen, we'll tell you that there are times you don't need it. There are times when you don't even want it. Metabolism without oxygen is called *anaerobic metabolism*. It's possible because there are other oxidizing agents besides free oxygen, although other oxidation/reduction reactions generate less energy. However, anaerobic metabolism creates many by-products of interest to bioengineers, such as lactic, butyric, succinic, or proprionic acids, and ethanol, butanol, or propanol. Fermentation is an example of anaerobic metabolism. Fermentation of grapes and other products produces ethyl alcohol. The presence of oxygen kills this reaction. Well, it doesn't kill it as much as it drives it to completion.

So, if you are trying to produce ethyl alcohol by the fermentation of grapes and oxygen enters the pictures, you get carbon dioxide and water instead of alcohol. The engineering of the fermentation reaction to produce alcohol requires methods to keep oxygen out, methods that have been in existence for millennia. A summary of the ways that cells generate energy is given in Figure 1-16.

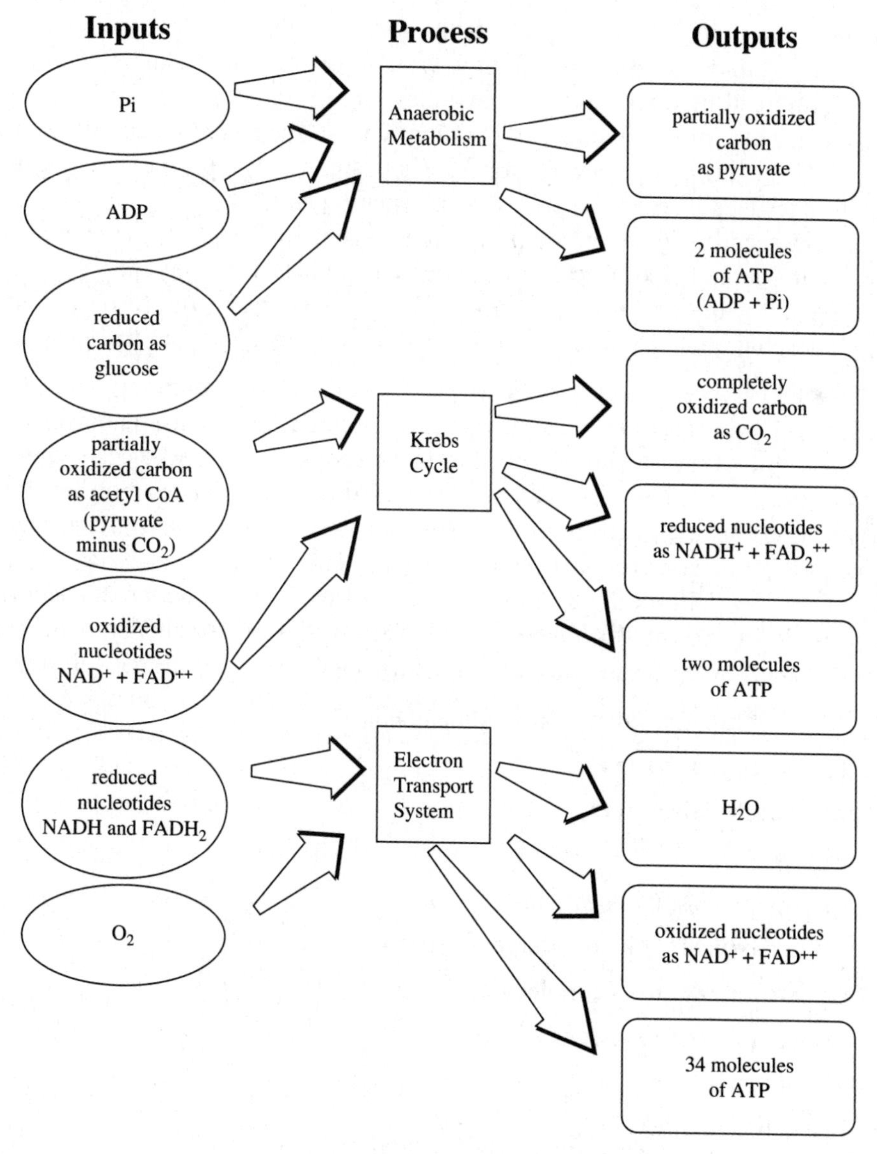

Figure 1-16 Summary of energy strategies of cells

Summary

Biological systems are composed of carbon, oxygen, hydrogen, and nitrogen. These molecules are used to build lipids, proteins, carbohydrates, and nucleic acids, the building blocks for cell structure and chemical reagents for cell function. An important function for lipids is the formation of the cell membrane. Molecules with a lipid tail and protein heads form a bilayered micelle that keeps the aqueous workings of the cell inside and protects the cell from the outside (also aqueous) environment. Lipids can also store energy. Proteins conduct the business of the cell by regulating cell function, and carbohydrates serve primarily as energy sources, although they also provide important building blocks. The receptors on the surface of the cell are primarily carbohydrates. They are highly specific and receive molecules destined to enter the cell. The information structure of the cell is found in nucleotides, consisting of only four different molecules. These are trapped inside the familiar double helix of deoxyribose nucleic acid (DNA). The energy for the cell comes from the oxidation of carbon. Energy is stored in the form of high-energy phosphate bonds coupled to a nucleic acid in an adenosine tri-phosphate molecule (ATP).

Quiz

1. Biomolecules:
   (a) behave differently than molecules outside of a living system.
   (b) are formed from a backbone of hydrogen.
   (c) are formed primarily from ionic bonds.
   (d) are of four basic types.
   (e) a and d are correct.
2. Lipids:
   (a) do not dissolve in water.
   (b) are typically found in large polymers.
   (c) are energy storage molecules.
   (e) are bioactive molecules.
   (f) all are correct.
   (g) a, b, and c are correct.

3. Carbohydrates:
   (a) are formed from C, H , O, and N.
   (b) would leave a carbon atom if atoms forming water were removed.
   (c) form readily accessible storage in the form of starch and glycogen.
   (d) include sugars.
   (e) c and d are correct.
   (f) b, c, and d are correct.
4. Proteins:
   (a) are composed of amino acids.
   (b) are energy storage molecules.
   (c) are structure molecules.
   (d) orchestrate the chemical reactions of the cell.
   (e) all are correct.
   (f) a, b, and c are correct.
   (g) a, c, and d are correct.
5. Peptide bonds:
   (a) are found in all biomolecules.
   (b) involve a link between a carbon and a nitrogen atom.
   (c) are ionic bonds.
   (d) result in a hydrophobic molecule.
6. Nucleic acids:
   (a) include organic bases.
   (b) include phosphorylated sugars.
   (c) are found in the nucleus.
   (d) contain the information structure of the cell.
   (e) all are correct.
   (f) a, c, and d are correct.

7. Hydrogen bonds:
   (a) are ionic bonds.
   (b) are covalent bonds.
   (c) are the bonds between the bases in DNA.
   (d) are important in explaining the behavior of water.
   (e) c and d are correct.
   (f) b and c are correct.
8. DNA:
   (a) typically consists of two strands.
   (b) strands are made of a backbone of organic bases.
   (c) phosphorylated sugars form the steps of the "staircase."
   (d) base pairs associate randomly through hydrogen bonds.
   (e) all are correct.
9. Polymers:
   (a) consist of linkage of small units known as monomers.
   (b) are the usual form of biomolecules.
   (c) may contain units of the same or different monomers.
   (d) the loss of one monomer may change the function of the polymer.
   (e) all are correct.
10. DNA:
   (a) forms a double helix.
   (b) always contains the same amount of adenine as guanine.
   (c) always contains the same amount of adenine as thymine.
   (e) consists of two strands that require a large amount of energy to pull apart.
   (f) a, b, and e are correct.
   (g) a and c are correct.

CHAPTER 2

Cell Structures and Cell Division

To engineer a living organism, you must

- Form a barrier between the organism and the organisms environment. The barrier must allow the cell to select which material may enter and which may not.
- Defend the organism against unhealthy conditions.
- Have an energy source for the organism.
- Control production of the biomolecules.
- Move molecules around the cell.
- Eliminate waste.
- Have a means of duplicating the organism.

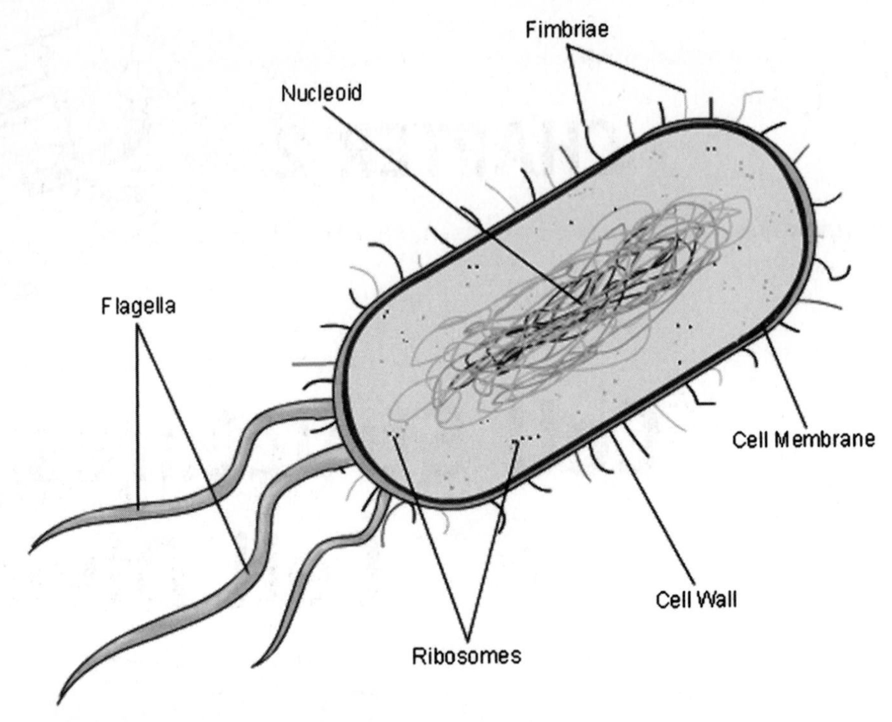

Figure 2-1 Prokaryotic cell

All cells have a lipid membrane and a watery, jelly-like substance on the inside, known as *cytoplasm.* The most primitive cells don't have interior compartments, other than a few vesicles for eliminating waste. These cells are known as *prokaryotic* cells, as shown in Figure 2-1. The "karyotic" portion is derived from a Greek word meaning "kernel." The prokaryotic cells are so named because they were derived before chromatic material was organized into a nucleus, that is, before cells had a kernel. Their nucleic acids are in the cytoplasm with everything else, where they form a dark area known as the *nucleoid.* These cells include unicellular bacteria.

Most cells are more advanced and are called *eukaryotic* because they have a nucleus and numerous functional compartments called *organelles.* (See Figure 2-2.) Only primitive bacteria are prokaryotic. Prokaryotic cells don't have the capabilities of eukaryotic cells, which can become highly specialized. However, you as the bioengineer are interested in prokaryotic cells because their simplicity makes them easier to deal with. But the differences between the prokaryotic cells and the eukaryotic cells can make a huge difference in the way they process genetic information, and you need to understand these differences.

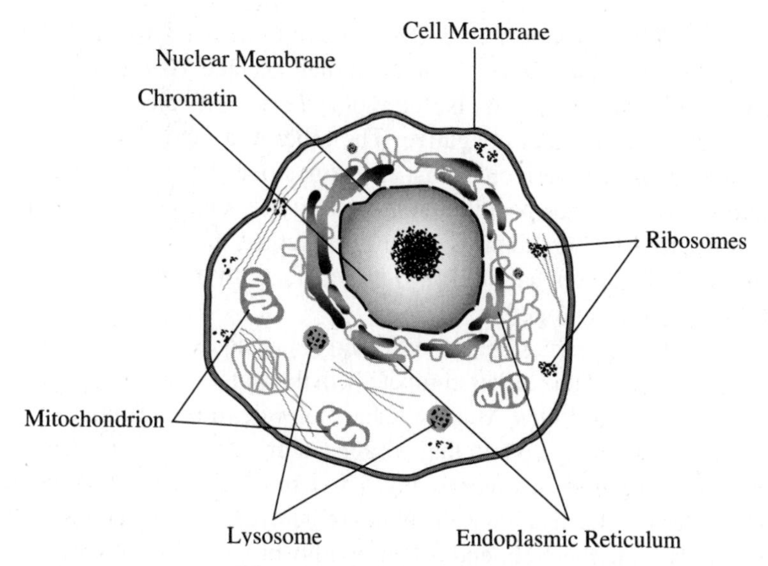

Figure 2-2 Eukaryotic cell

Most of the structures discussed next are distinguishable only in the eukaryotic cell, but the general functions described are also found in the cytoplasm of the primitive prokaryotic cell.

The Barrier: Cell Membranes

All cells possess a cell membrane. I am sure that you know what a cell is...basic unit of an organism, right? Actually, in a multicellular organism like us, each cell is an oil vesicle bathed in ocean water or water with a concentration of sodium chloride (salt) identical to ocean water. Inside of this oil vesicle exists a marvelous chemical system that allows each cell to grow, divide, fight off intruders, take in important reagents, and excrete other materials that are either waste products or important products of this particular cell's metabolism. Unicellular organisms are not so lucky. In many cases, they surround themselves with a capsule or *cell wall* to protect themselves from a hostile environment.

To understand the forces that create this oily ball, envision a drop of oil in a container of water. The oil drops stay intact. Now add another drop of oil to the container. And then another. The droplets will coalesce, as if they are seeking each other's company and are avoiding the water. Chemicals that behave this way are called *hydrophobic*, or water-hating. The underlying reason for their behavior in water is that they have no electrical charge at their surface. As discussed in Chapter 1, water molecules are polar, that is, the molecule exhibits electrical charges. One end, the oxygen end, is electronegative. The other end (actually two ends), composed of the hydrogen molecules, is electropositive. The water molecules are attracted to each other with positive ends seeking the negative ends of other molecules. As the partially-charged ends try to get together, they force out anything that gets in the way, like an uncharged (oily) molecule, with the net effect that the oily molecules tend to clump together.

Molecules that bear charges behave differently in an aqueous solution. The charged areas (poles) of the other molecules compete with the water molecules for the opposite charges on the water molecules. As a result, the molecules migrate until they are evenly dispersed. From our vantage point, they disappear, or dissolve in the water. This type of molecule is *hydrophilic*, or water-loving. These molecules have polar ends.

The molecules that form cellular membranes are hybrids—one end is hydrophobic or oily (non-polar) and the other end is hydrophilic, or water-loving (polar). The molecules line up in two layers because their non-polar tails seek out one another or, conceptually, are pushed together by water. The lipid tails cluster together with the tails of other molecules and induce the formation of a second layer. Polar heads composed of sugar protrude on either side of this double layer. Then, depending on their number and size, they curl around to form a hollow ball. The ball is a hollow micelle. Actually, any collection of these molecules in an aqueous solution will behave the same way—line up tail-to-tail and with hydrophilic ends composed of polar sugars, projecting outside of the double-layered membrane. The micelle encloses a watery interior. This interior is isolated from the rest of the environment by its oily barrier. The chemical business of the cell can now be conducted in the privacy of the watery interior, safe from disruptive environmental forces. See Figure 2-3.

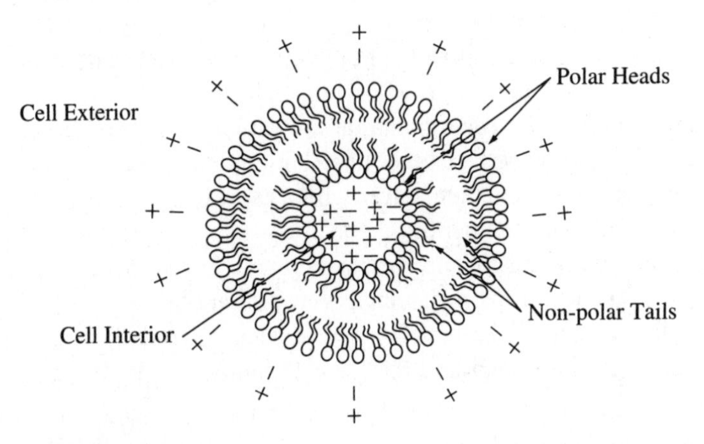

Figure 2-3 The cell membrane and the cell as a micelle

This is a great system because in order to dissolve in blood, a molecule must be water soluble, but in order to penetrate the cell, a molecule must be lipid soluble. The lipid barrier keeps out a lot of nasty stuff. The membrane allows some things to pass, such as small molecules of water or CO_2. However, most bioactive molecules are charged and therefore blocked and need permission somehow to enter the cell. The way the permission is granted is through the existence of specific receptors that appear on the surface of the cell. The receptors are typically carbohydrates attached to a protein stem, and the membrane behaves as a viscous semi-liquid. The protein molecules that contain receptors float up through the membrane and stick their receptors above the surface. These receptors are specific to unique molecules and provide the entrance to the cell because the attached molecules can move through the protein into the cell or be actively carried in. Sometimes binding the molecule negates the charge, at which point the entire neutral molecule passes through readily. We will see that the type and amount of different receptors that are produced by the cell and that are transported to the surface, are the primary mechanism by which the cell controls what and how much of what enters.

Consider insulin, which regulates sugar metabolism within the cell. Insulin must attach to a receptor at the cell surface to be active. Therefore, the body can control sugar metabolism at two points—first, the amount of insulin that is produced and second, the amount that attaches to the cell membrane (proportionate to the insulin receptors that are on the surface of the cell).

Most bacteria, plants, and fungi have an additional protective layer called a *cell wall* made of a protein-carbohydrate complex. The cell wall functions to protect the organism from environments with a different salt concentration than the cytoplasm. Without the cell wall, water would attempt to balance the salt concentration between the inside and the outside of the cell, and the cell would either shrink or burst.

The Genetic Material: The Nucleus

Eukaryotic cells have a double-layered inner membrane that protects the chromatin, the genetic material. This membrane is called the *nuclear membrane* and the entire structure is known as the cell *nucleus.* The nucleus provides a housing for the processes involving DNA. The interior of the nuclear membrane is connected to the fibrous lamina, which forms thin sheets near the membrane, and each chromosome is accorded a specific position where it is attached to the lamina. The nucleus also contains a darkened area, the *nucleolus*, devoted to the manufacture of the RNA found in ribosomes.

Prokaryotic cells usually have only one chromosome, which is located in a darkened area called the *nucleoid.* There is no protection for this chromosome from the general milieu of the cytoplasm.

The Energy System: Mitochondria

You learned in Chapter 1 that the energy of the cell is provided by an oxidation/reduction reaction, and the most efficient of this reaction utilizes free oxygen. Eukaryotic cells have a unique organelle, called a *mitochondrium*, in which the energy storage molecule (ATP) is produced using oxygen in a process known as *oxidative metabolism.* The cliché is that mitochondria are the "powerhouses" of the cell. See Figure 2-4.

Mitochondria possess an inner membrane that is repeatedly folded inside the organelle. The folds are known as *cistae*. Like any surface folded inside a given space, the folds provide a large surface area, and it is upon this surface area that the oxidative metabolism occurs. The enzymes necessary for the oxidation/reduction reaction are there. We will talk about enzymes in detail in Chapter 8.

Mitochondria are interesting because they have some features that make them seem like their own little life-form. Perhaps they were once independent organisms that invaded other cells early in the evolution of life. Mitochondria contain their own DNA (called *mitochondrial DNA*). So, in addition to an energy source, mitochondria have their own information system that codes for proteins within the mitochondria. Furthermore, mitochondria can duplicate themselves. It is easy to envision them as independent organisms.

Mitochondrial DNA divides separately and is independent of the other cellular genetic material. As a result, there is no shuffling of DNA in the creation of an embryo. Mitochondrial DNA is passed intact from mother into progeny within the mitochondria of the egg, and sperm have no mitochondria. In Chapter 7, the use of mitochondrial DNA to trace ancestral genetic patterns will be discussed in detail.

Prokaryotic cells, without the organelles known as mitochondria, are limited to the inside of the cell membrane to provide a substrate for oxidative metabolism.

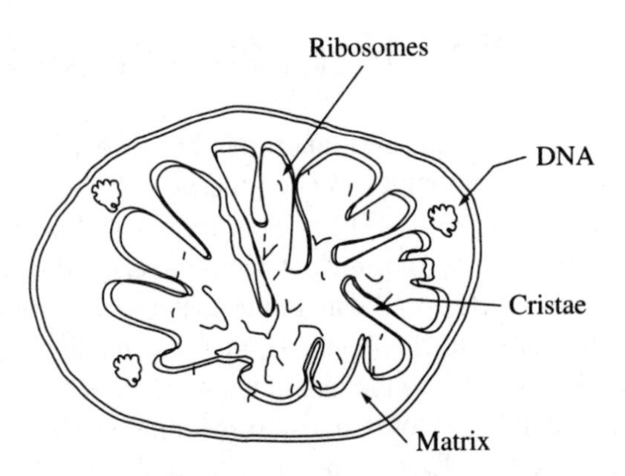

Figure 2-4 Mitochondria

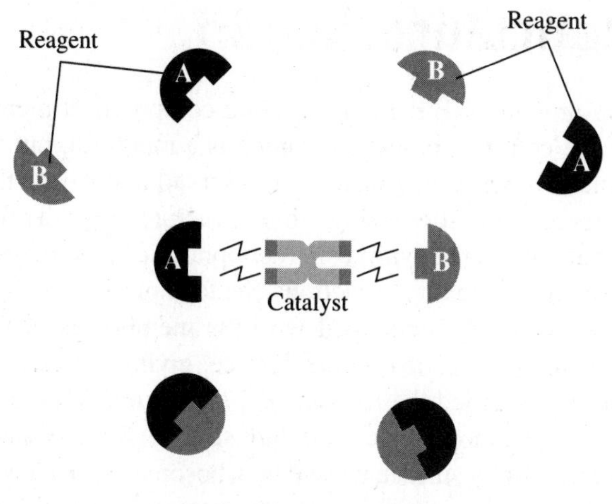

Figure 2-5 Catalysts

Control of Cell Chemistry

DNA dictates the production of specific proteins, and proteins make everything else happen. Proteins that are involved in regulating cell function are called *enzymes*, which may break down other molecules. Also, as *catalysts* (shown in Figure 2-5), they may drive a reaction that produces other molecules. Briefly, catalysts are necessary for a given reaction but are not direct participants in the reaction. They may bind other molecules together in just the right configuration so that the molecules that are held in place by the enzyme will react, or they may break off strategic parts of a molecule to make a reaction move forward.

Ribosomes

Ribosomes are floating protein factories. You will learn how ribosomes work in Chapter 3, but a brief description is that the DNA calls for the production of a particular protein by issuing an order in the form of messenger RNA (mRNA). The ribosomes lock onto the mRNA, translate the order (sequence of bases), and produce the protein. Both eukaryotic and prokaryotic cells contain ribosomes, although the ribosomes found in prokaryotic cells are significantly smaller than those found in eukaryotic cells.

Endoplasmic Reticulum

The *endoplasmic reticulum* (E.R.) is a structure composed of membranes in folds on the inside of the cell membrane. The name is a mouthful, but know that these structures were named when they were first observed and before the function was known. Early researchers observing this structure saw a folded structure (reticulum) protruding into the cytoplasm (endoplasmic). The folded structure provides surface area where chemical reactions can take place.

Ribosomes are frequently associated with the membranes of the endoplasmic reticulum. They make little knobs on the surface, giving the E.R. a rough appearance. You will hear this called the *rough endoplasmic reticulum*. Protein synthesis is intense on the rough endoplasmic reticulum surface. Some areas of the E.R. are smooth, lacking the knobby surface caused by ribosomes. The *smooth endoplasmic reticulum* is devoted to production, processing, and, in some cases, destruction of lipids.

Golgi Apparatus

The *Golgi apparatus* serves to sequester proteins. How did the name come about? Well, it was discovered by Dr. Camille Golgi in 1898. This organelle is similar to the E.R. and has small internal compartments or sacs called *cisternae*. The Golgi apparatus receives proteins from the E.R. and extrudes them in the direction of the cell membrane. Also, as discussed in Chapter 8, proteins undergo processing after they are produced. The processing involves reconfiguration before they are secreted to the exterior of the cell. Much of this type of processing occurs within the Golgi apparatus.

Transporting Material and Waste Elimination

The issue of transporting material around the cell is somewhat problematic. Remember that the inside of the cell is friendly to water-soluble or hydrophilic molecules and not compatible with water-insoluble or hydrophobic molecules. However, inside the cell, molecules that are destined for the cell membrane may be hydrophobic and would coalesce within the aqueous cytoplasm. You would also need to protect the cell from certain enzymes produced by the cell, enzymes that are dedicated to waste disposal. Such enzymes might digest essential cell structures. Eukaryotic cells have developed clever ways to transport molecules to their point of action.

One method of moving material around is within small transport compartments surrounded by a hydrophobic (lipid) membrane. These compartments are called *vacuoles*. The vacuoles can transport material to and from the cell surface in a mode isolated from the rest of the cell interior.

We will see in Chapter 8 how after proteins are produced, they may be folded to expose their hydrophilic side, transported through the aqueous cell interior, and then refolded to expose a hydrophobic surface. Alternatively, the hydrophilic surfaces may be capped with small molecules called *chaperone modulators* converting them to hydrophobic surfaces. The hydrophobic surface allows them to float through the cell membrane. Some of these proteins that are produced on the surface of the endoplasmic reticulum migrate into the reticulum folds where they are ideally positioned to float into the cell membrane, perhaps to be excreted.

In the course of conducting their business, cells produce waste. For example, the complete oxidation of carbon results in carbon dioxide and water. Both of these diffuse freely into and out of the cell. There are other metabolic processes that produce excess organic molecules that the cell must eliminate. An example is ethyl alcohol, produced as a by-product of anaerobic metabolism. Such materials may be transported to the cell surface surrounded by a protective vacuole, which fuses with the cell membrane and extrudes its contents.

Eukaryotic cells contain packages, called *lysosomes*, of digestive enzymes known as *lysozymes*. Lysosomes are formed when pieces of the Golgi apparatus are pinched off as a sac containing the lysozymes. Lysozymes gain access to the material they need to digest when the lysosomes fuse with vacuoles containing waste products, as shown in Figure 2-6.

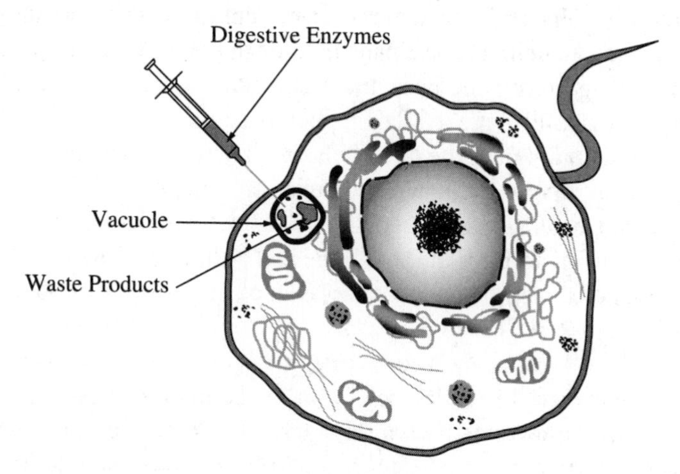

Figure 2-6 Lysozymes are dumped into a vacuole to digest waste molecules.

Cell Duplication—Mitosis

If you were gazing into a microscope watching a cell divide, here is what you would see. First, you would see the material in the nucleus become very dense. You would be able to discern distinct, separate structures—the chromosomes. You know that in human cells there are 23 pairs of chromosomes, one of each pair from each parent, a total of 46 chromosomes. By Jove, all 46 would line up in the center of the cell, north to south. At this point, you would see that each of the 46 chromosomes is actually an identical pair, hooked in the center with what looks like a button. At opposite ends of the cell, east and west, you would notice two dense spots, each of which appears to be generating a semi-circular spoke of lines. Each line connects to a chromosome pair at the button. There is one line from the east spot and one line from the west spot attached to each pair of chromosomes. Then, as if the dense spots at the poles of the cell were acting as reels, pulling in the lines, the chromosome pairs split and are dragged to opposite sides of the cell, east and west. At the same time, the cell splits, north to south. This is an amazing sight.

The events continue as follows. Understand that to duplicate a cell, the essential task is to copy the DNA. The instructions for making everything are encoded in the DNA. Remember that DNA is composed of a double strand. The outside of the strand is a scaffolding of sugars held together by covalent phosphate bonds. Inside are organic bases, found in pairs, linked by hydrogen bonds. The strategy for copying the information stored in one cell's chromosomes and passing the information to a new cell in its entirety is very simple: unwind the DNA and break apart the hydrogen bonds that hold the base pairs together. At this point, you have two single strands with an organic base protruding from the sugar backbone. Remember that a given base always pairs with the same partner, adenine with thymine and cytosine with guanine (complementary base pairing). Also remember that a *nucleotide* is a molecule containing an organic base and a phosphorylated sugar. Theoretically, if the appropriate nucleotides are available, each base on each strand will pick up its matched nucleotide pair out of the nucleoplasm. (There is a specific and very important enzyme, DNA *polymerase*, which makes this happen.) The sugars and the phosphates of the new, juxtapositioned nucleotide will join together with their phosphodiester linkage to form the backbone and voilà! You now have two identical chromosomes where once there was one...except, it's a little more involved than just that. See Figure 2-7.

A more involved look at the process begins with chromosome pairs scattered throughout the nucleus. The resting cell is said to be in *interphase*. In a resting cell, the genetic material known as *chromatin* appears as a tangled mass and is not as dense as it becomes when it is recognizable as *chromosomes*. (Chromatin was named because it stained darkly with colorful dyes, long before anyone knew its

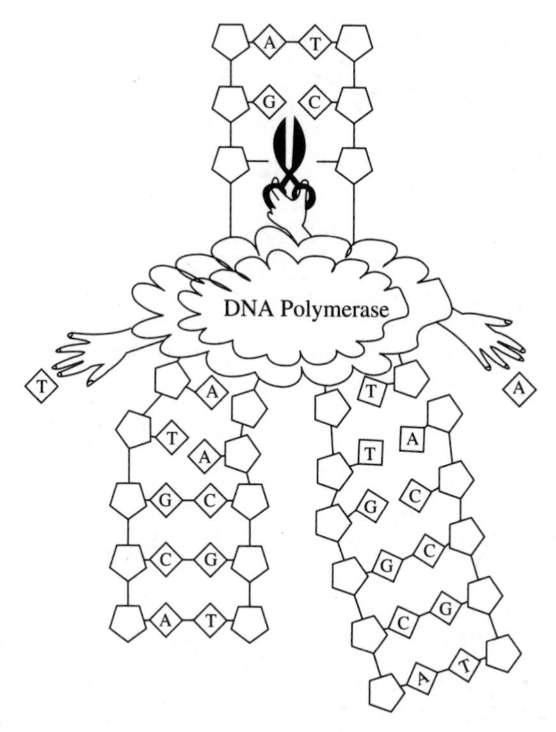

Figure 2-7 The process of duplicating DNA

function.) The human cell has 23 pairs of chromosomes. Twenty-two of these pairs are called *autosomal chromosomes* and are the same for males and females. The twenty-third pair of chromosomes is the *sex chromosome* pair and is different for males and females. Females have two large chromosomes, called X, and males have one large X and one smaller Y chromosome. Chromosomes of a pair are of same type, but the chromosomes from different parents contain different information. In that sense, the chromosome pairs are not identical.

Mitosis, shown in Figure 2-8, is the process by which the cell duplicates itself, and it occurs in five distinct phases. When events are set in motion to divide the cell, each member of each chromosome pair is duplicated. This occurs in the *prophase* cycle of mitosis. All 46 chromosomes are duplicated. These duplications are truly identical and duplicates are called *sister chromatids*. Temporarily, the cell has 92 chromosomes.

For the cell to begin the process of duplicating the chromosomes, unzipping the chromosome is a bit of a challenge. Most of us think of chromosomes as long, winding staircases, because that is how we have always seen it depicted. Actually, even in a resting cell, the DNA material is wound very, very tightly, staircase on top of staircase. Consider the fact that the DNA inside a single cell, if laid end to end,

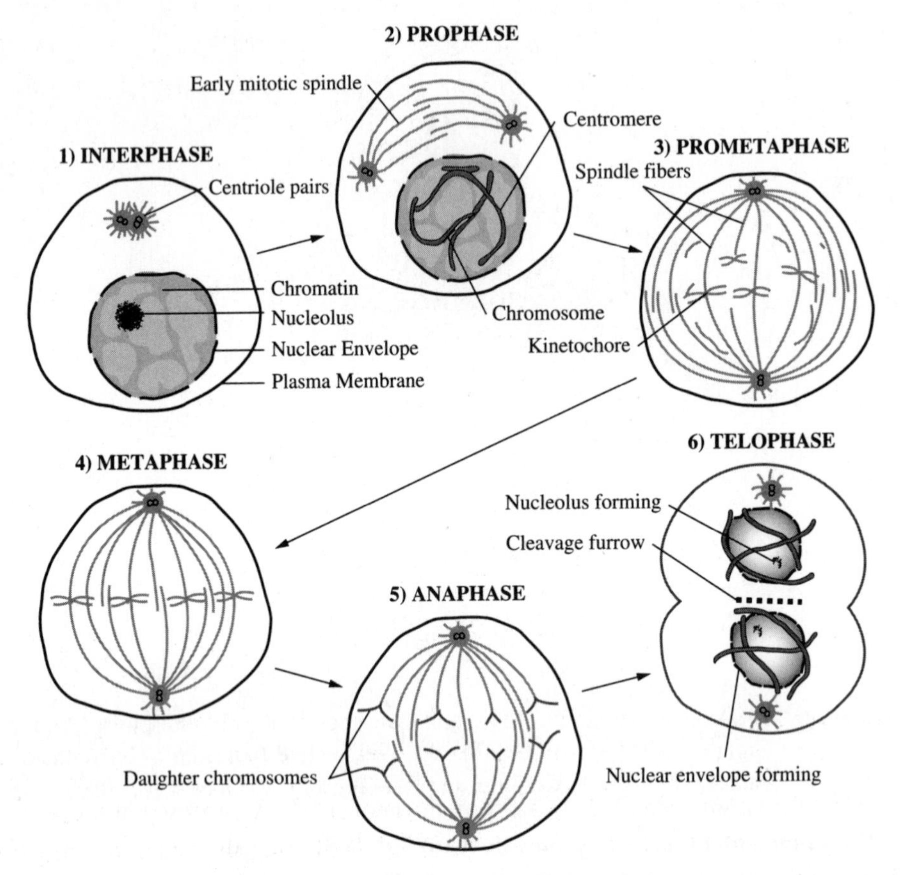

Figure 2-8 Mitosis

would be over an inch long! Think of how clever the packaging must be to pack that much material inside of a microscopic cell. Instead of picturing a staircase, visualize the double-stranded DNA as a piece of twine. The DNA winds around itself like many strands of twine forming a rope. Then, in eukaryotic cells, the DNA rope winds around proteins balls called *histones*. Now the whole complex twists to form flat discs. See Figure 2-9.

Before the DNA can be replicated, it must be "unwound" by an enzyme known as a *helicase*. Then the hydrogen bonds are broken, the stands are pulled apart, new strands are formed, and the material condenses into two separate chromosomes.

Cell division is tightly controlled. Certain cells divide frequently, such as bone marrow cells; certain cells divide primarily in growing or injured organs, such as bone cells; and certain cells divide rarely, if at all, such as mature nerve cells. Uncontrolled growth is the hallmark of cancer. The division process requires something to kick it off, and this "something" is called an *activation factor*. An example of an activation factor is the human growth factor.

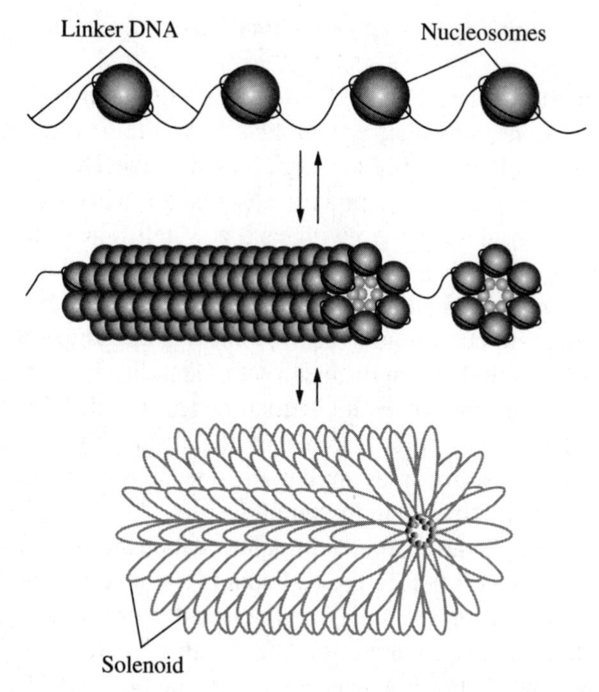

Figure 2-9 DNA wound around histones and then stacked into chromosomes (Based on Figure 3-1 of *Schaum's Outlines Molecular and Cell Biology* by William D. Stansfield, Jaime S. Colome, and Raul J. Cano, McGraw-Hill, 1996)

The way activation factors usually work is by altering the shape of the DNA, allowing the enzyme that performs the division of the chromosomes, DNA polymerase, to bind. The DNA polymerase first finds a specific shape on the DNA. This shape, a loop, is required before the polymerase can bind to the DNA and do its thing. This looped configuration is assumed when the activation factor binds to the DNA. The area where the DNA polymerase initiates DNA replication is called the *origin of replication*. You should know that the DNA polymerase requires a *primer*, composed of specialized RNA molecules before the duplication can begin. Once the DNA polymerase finds a binding site, it chugs up the chain, separating the strands and building new ones by adding complementary base pairs to the single strands of DNA.

One of the most important jobs of the DNA polymerase is to correct errors in the newly synthesized DNA. When an incorrect base pair is recognized, the DNA polymerase reverses its direction by one base pair of DNA. The incorrect base pair is excised and the correct one inserted in an activity known as *proofreading*. As a result, an error in DNA replication is propagated only about one in each 10–100 billion base pairs. However, because cells divide so frequently, mistakes are inevitably transmitted. Many of the mistakes result in cells that are not viable. If a mutated cell does survive, the immune system has cells that ferret out and destroy these abnormal cells.

Now let's return to the business of dividing the cell. The first phase of mitosis is completed when the chromosomes are duplicated. In the second phase of mitosis, *prometaphase*, the nuclear membrane dissolves and the chromosomes begin to move to the center of the cell. This process ensures that when the cell divides, one of each set of chromosomes must go with each new cell. The sister chromatids are attached at points called *kinetochores* through a central dense area called a *centromere*. This is what looked like a button to you when you first saw this event. There are two peripheral hook points (*centrosomes*) at opposite sides of the cell, as you observed. Before cell division occurs, a set of spindles is formed originating at the hook points and attaching to each kinetochore from each side of the cell. Each spindle fiber hooks to one chromosome. In the third phase of mitosis, called *metaphase*, each of the centrosomes applies equal tension to one of each chromatid pair. As a result, the pairs line up in the middle.

The spindles pull duplicate chromosomes, now called *daughter chromosomes*, to the opposite side of the cell before the cell splits. The phase wherein the chromosomes move to different areas is called *anaphase*. If everything goes according to plan, the kinetochores separate and one of each pair of sister chromatids is pulled to each side of the cell. If chromosomes do not line up properly, the process breaks down.

After the chromosomes have been distributed to different areas of the parent cell, the cell splits in two, a process called *cytokinesis*. The fourth and final stage of mitosis is called *telophase*. Cells produced in this way are *somatic cells* and include every type of cell in your body, except the cells that become sperm and egg. The cells that will contribute to the formation of a new organism, known as *germ cells*, are created by a process called *meiosis*, discussed next.

Again, mitosis usually goes without a hitch. However, there are literally millions of cells dividing in every human body at any given time. Occasionally, a cell ends up with too much genetic material and other cell is shorted. Usually, such cells do not survive or are eliminated by the immune system.

Sexual Reproduction—Meiosis

Mitosis is the process whereby a somatic cell duplicates itself. *Meiosis*, shown in Figures 2-10 and 2-11, is the process to prepare *germ cells* or *gametes*. A germ cell contains only one chromosome from each parent and is designed to fuse with a germ cell from another individual to produce a genetically unique offspring. Cells that contain only one chromosome from each pair are said to be *haploid* (one copy),

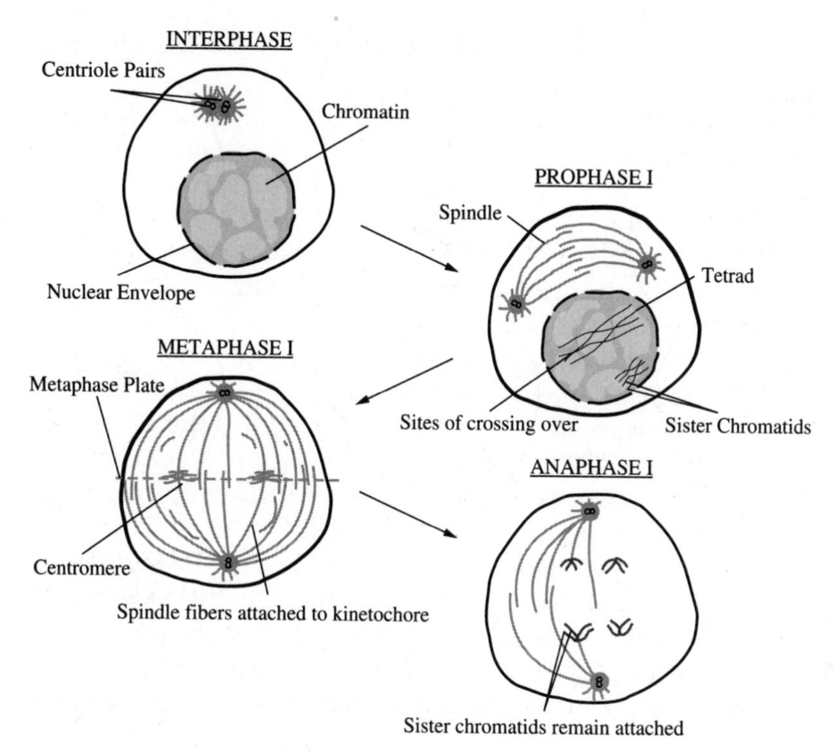

Figure 2-10 Meiosis I

whereas normal cells are *diploid* (two copies). Eggs are germ cells from females and sperm are germ cells from males. On the surface, producing a germ cell would appear to be easy to do: simply pair up the chromosomes, divide the cell, and send one of each chromosome pair to each new cell. You would probably want to ensure some kind of shuffling so that you did not recreate the parental germ cells.

Actually, meiosis contains several steps that seem to be unnecessary. In a nutshell, the precursors to the gametes duplicate their chromosomes once and divide twice. The process results in four haploid cells. Why would this be advantageous? Why not just split the cell without duplicating the chromosomes?

As we describe sexual reproduction, it will probably strike you that the process is optimal for reshuffling the genes, with the effect of creating genetic diversity. At the gross level, genetic diversity is ensured by the fact that our chromosomes occur in pairs, one from each parent. So, your genome includes a gene to produce, theoretically, a crucial enzyme that breaks down a certain toxic by-product of metabolism. You should have two copies of this gene. If one is defective, the other gene may be able to compensate; the system is robust in that sense. Meiosis allows chromosomes to exchange genetic material, creating additional genetic diversity

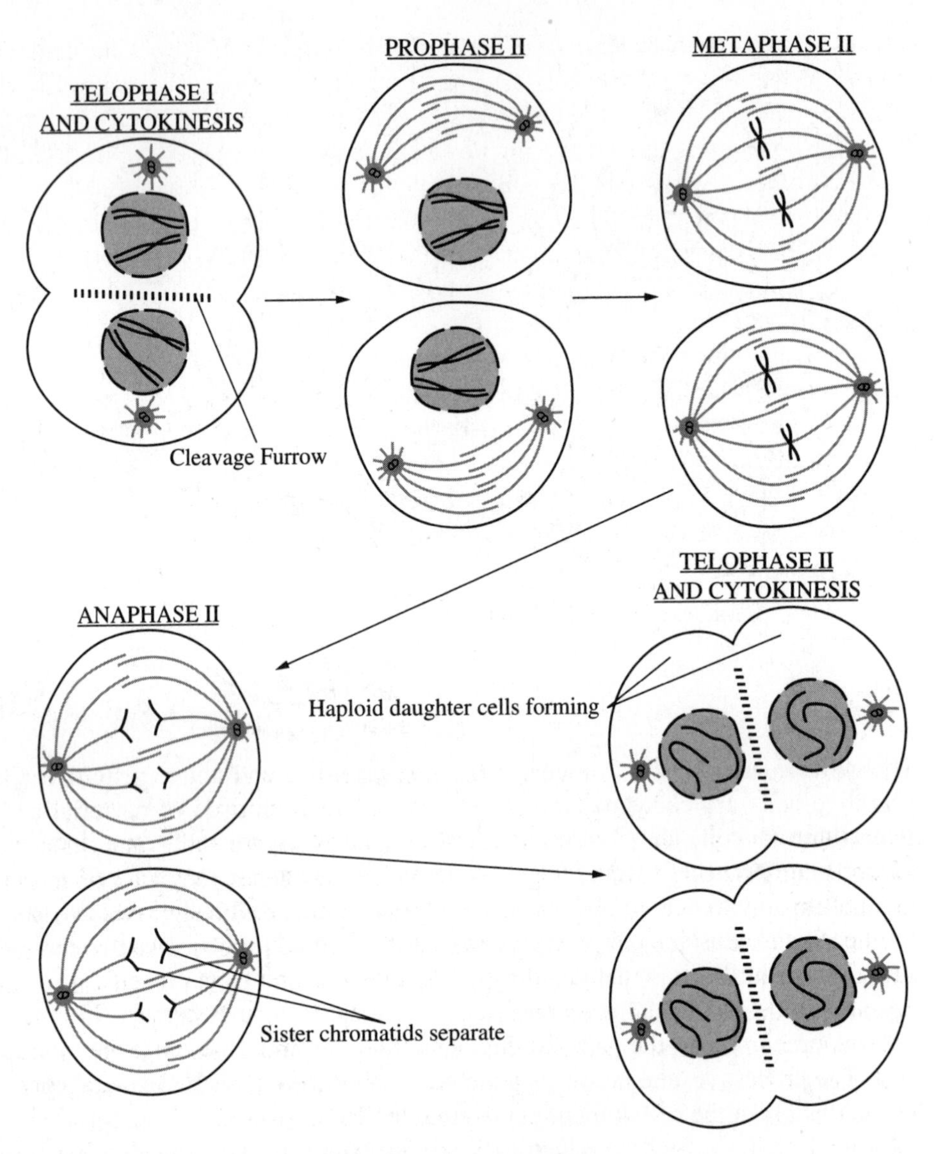

Figure 2-11 Meiosis II

within a given population. This increased diversity results in populations more adaptable to changes in their environment than populations with less genetic diversity.

The steps in meiosis are as follows. When the chromosomes divide, they line up in the center of the cell, all in a nice row just like in mitosis, except that the chromosome pairs align. As a result, the chromosomes are four abreast, and this structure is called a *tetrad*. For the inner two chromosomes, one of the chromosomes from one parent is adjacent to the same chromosome from the other parent. Whereas in

human mitosis, you have 46 rows of sister chromatid pairs lining up in the center of the cell, in meiosis, you have only 23. This configuration is called a *synapsis*. If you were observing this four-abreast configuration, you would see the limbs of the inside chromosomes begin to intertwine, as if in an ancient embrace. Then portions of the chromosomes actually change places.

Clearly, genes that are far apart on the chromosome are more likely to become separated than genes that are close together. When we discuss the research that determines the location of a given gene on a chromosome, you will learn that one way of determining how far apart genes are on a chromosome is by how often they get separating during this mix-and-match exercise. As a result of pieces of the DNA being cut out of one chromosome and pasted into another, the four haploid cells from a single precursor cell are all different, that is, genetically diverse! For an organism turning out offspring into a hostile world, you can see how it would be desirable to have each youngster a little different, hoping for that slight advantage that would enable at least some of the offspring to survive.

Summary

The cellular world can be divided two parts, primitive and not so primitive. The primitive cells are called prokaryotic cells and are comprised of bacterial cells. Everything else, including yeast, fungi, and human cells, are eukaryotic cells. Prokaryotic cells are very simple: they have relatively few genes, they have no internal organelles, and they don't even have a nuclear membrane. Bioengineers love them because they are so easy to grow and maintain. Eukaryotic cells are more complex and capable of more complicated functions. Prokaryotic cells conduct all of the business of the cell in their cytoplasm.

Both types of cells are surrounded by a double-layer cell membrane. The interior surfaces of the membrane are composed of hydrophobic molecules that keep environmental water out and cellular water in, respectively. The exterior of the membrane is composed of hydrophilic sugar molecules. Literally floating in the membrane are proteins with carbohydrate heads. The proteins float up to the outer surface and stick their heads out into the environment. These heads serve as specific receptors that will bind to certain molecules that the cell is, essentially, looking for. Once the molecule is found and bound to the receptor, it can be taken into the cell.

Eukaryotic cells have compartments, called *organelles*, where specialized functions are conducted. The genetic material is in the cell nucleus, surrounded by a cell membrane. The metabolic processes of the cell are conducted in the mysterious little mitochondria. The Golgi apparatus stores proteins. The endoplasmic reticulum is a folded surface that provides surface area for the manufacture of proteins and other cellular processes. Ribosomes, the protein-manufacturing plant of the cell, are frequently located on the endoplasmic reticulum.

Cell division of somatic cells (all cells except those that produce sperm and egg) is called *mitosis*. During mitosis, the chromosomes are duplicated and split between each of the new cells. The process is dependent on an enzyme called DNA polymerase. The DNA polymerase looks for a specific looped shape on the chromosome to attach to and begin splitting the DNA molecule apart. Factors that control cell divisions act by creating, or alternatively, preventing, the formation of this looped structure.

Cell division of germ cells occurs by a process known as *meiosis*. In meiosis, the chromosomes duplicate once and the cell divides twice. The process is designed such that paired chromosomes, one from each parent, exchange genetic material before they are sorted into the germ cells, which have half the normal amount of chromosomes. This condition is known as *haploid* (one chromosome per type) as compared to *diploid* (two chromosomes per type).

Quiz

1. Prokaryotic cells differ from eukaryotic cells by:
   (a) a lack of a cell membrane.
   (b) a lack of a nuclear membrane.
   (c) a lack of an ability to perform oxidative metabolism.
   (d) a lack of an ability to secrete waste products.
   (e) b and c are correct.
2. The cell membrane:
   (a) freely admits water soluble molecules.
   (b) excludes lipids.
   (c) prevents the entrance and egress of water.
   (d) exhibits specific receptors on its surface.
   (e) b and d are correct.

3. Proteins:
   (a) govern chemical reactions within the cell.
   (b) are produced by ribosomes.
   (c) may reside within the cell membrane.
   (d) may be hydrophilic or hydrophobic, depending on how they are folded.
   (e) all are correct.
4. The endoplasmic reticulum:
   (a) is the site of oxidative metabolism.
   (b) increases the surface area available for chemical reactions.
   (c) is created by folds within the nuclear membrane.
   (d) is the area in which wastes are digested by the cell.
   (e) b and d are correct.
5. Mitochondria:
   (a) are frequently found on the surface of the endoplasmic reticulum.
   (b) are the site of oxidative metabolism.
   (c) contain distinct DNA structures.
   (d) can replicate themselves.
   (e) all are correct.
   (f) b, c, and d are correct.
6. Meiosis:
   (a) is carefully conducted to preserve the integrity of individual chromosomes.
   (b) creates four haploid cell from one precursor diploid cell.
   (c) provides exchange of genetic material between sister chromatids.
   (d) occurs only in prokaryotic cells.
   (e) b and c are correct.

7. Mitosis:
    (a) is the process by which cells reproduce themselves.
    (b) allows exchange of genetic material between members of a chromosome pair.
    (c) duplicates the chromosomes once and divides the cell twice.
    (d) is mistake-proof.
    (e) all are correct.
8. DNA:
    (a) occurs as 46 chromosomes in a normal human cell.
    (b) occurs as 92 chromosomes in a cell preparing to divide.
    (c) is loosely strung like long ropes throughout the nucleus.
    (d) is wound around protein balls in a resting cell.
    (e) a, b, and d are correct.
    (f) a, b, and c are correct.
9. Duplication of DNA:
    (a) requires an enzyme called DNA polymerase.
    (b) begins with the binding of an activation factor.
    (c) involves binding the DNA polymerase onto the DNA.
    (d) requires breaking the hydrogen bonds between base pairs.
    (e) all are correct.
10. Cell replication:
    (a) requires that all cell structures be carefully copied.
    (b) must be carefully controlled to prevent cancer.
    (c) sometimes results in abnormal cells.
    (d) occurs infrequently in adult humans.
    (e) a, b, and c are correct.
    (f) b and c are correct.

CHAPTER 3

Information Methods of a Cell

A sequence of organic bases, using only four different molecules, provides a code that orchestrates the business of life—all the way from the formation of cells, through conduct of cellular processes, reproduction, and even death. This is possible because the organic bases are instructions that cause the production of proteins—the agents of change. This chapter describes the mechanisms by which the code on the DNA results in the production of proteins and thereby in the activity we call life. A recent NOVA documentary on PBS ("Cracking the Code of Life," first aired April 17, 2001) explained the way that information is communicated within a cell with an analogy to cooking. This excellent analogy bears repeating, and, as the story unfolds, you will see the major players tied to a scene in a kitchen. It starts with the recipe—the DNA code.

History

In less than one hundred years, we have progressed from suspecting that DNA is involved in transmitting hereditary information to actively engineering living things through manipulating the information contained in DNA. The first insight was that cells could "talk" to one another, apparently through exchanging molecules. A landmark experiment was performed by Oswald Avery in the 1940s using *virulent* (disease-causing) and *avirulent* (unable to cause disease) streptococcal bacterial. The two types of strep differ in the capsule that covers the bacillus. Virulent bacteria have a rough capsule, whereas avirulent bacteria have a smooth capsule. The virulent strep did not cause infection if the bacteria were killed prior to being administered to mice. Living avirulent strep was unable to cause infection. However, if live avirulent bacteria were administered with dead virulent bacteria, the mice became infected. How had the avirulent bacteria learned to cause infection from their dead cousins? Furthermore, the bacteria isolated from the blood of the infected mice had rough capsules. Apparently, they were able to change their physical appearance based on what they had learned from the dead bacteria. The researchers concluded that the dead bacteria contained information that was being collected and expressed by the live bacteria, as shown in Figure 3-1. This information must be retained in the molecular structure of the cell. Later experimenters isolated the material that carried the "rough capsule" information and demonstrated that it was DNA.

In spite of the accumulating evidence, researchers of the early 20th century continued to debate the fundamental information methods of the cell. They resisted the concept that the code for all of the proteins of the cell could be carried on DNA. After all, DNA's variability is limited to only four organic bases. Contrasted with proteins, which are composed of 20 amino acids, this limitation seemed insurmountable. It just seemed more probable that the secret of life lay with proteins.

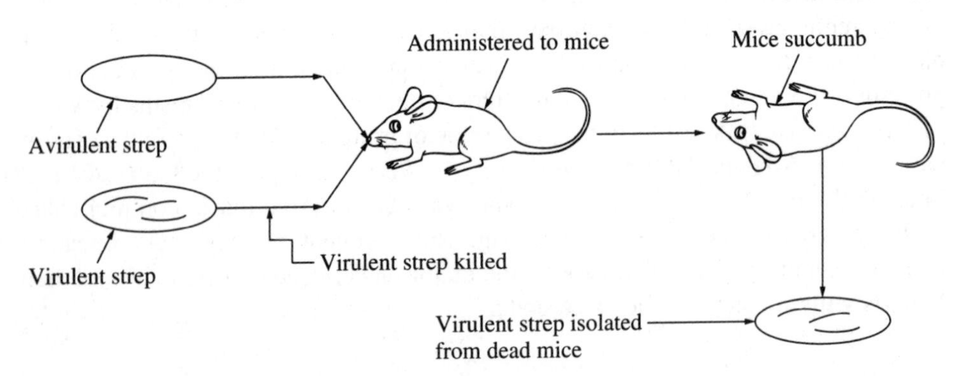

Figure 3-1 Transformation of avirulent streptococcal bacteria

Early researchers discovered another feature of DNA that, at first blush, seemed to limit the variability in the molecule. Less variability, less information, or so they thought. In any DNA sample from any organism, the amount of adenine was observed to be equal to the amount of thymidine, and the amount of guanine was identical to the amount of cytosine. These associations became known as"Charkoff's Principle." The association of adenosine with thymidine, and guanine with cytosine is the basis of *base pairing*. Recall from Chapter 2 that the pairing of these organic bases makes possible the duplication of DNA. When the DNA splits, each adenine picks up a thymine, each thymine picks up an adenine, each guanine picks up a cytosine, and each cytosine picks up a guanine. The result is two identical DNA molecules where once there was one. We will see later how base pairing makes possible the production of protein.

Once the evidence that DNA carried genetic information became irrefutable, the challenge became to understand how the components of DNA fit together to form a chromosome in such a way that the structure can be duplicated with fidelity. The race to understand the structure of DNA in the first half of the 1900s was similar in intensity to the race that occurred in the last half of the 1900s to decipher the DNA code. James Watson and Francis Crick finalized their model of the structure of deoxyribonucleic acid on the morning of February 28, 1953. The concept of a "double helix" that can "unzip" to make copies of itself provided final evidence that DNA carries our hereditary information. A significant insight came when they realized that an adenine-thymine pair held together by two hydrogen bonds is identical in shape to a guanine-cytosine pair. These pairs of bases could thus serve as the steps on a nice, uniform spiral staircase—DNA.

The significance of the predictable base pairing was not lost on Watson and Crick. James Watson said mildly, "It has not escaped our notice that the specific pairing we have postulated immediately suggests a possible copying mechanism for the genetic material."

Forty-eight years later, in February 2001, the International Human Genome Sequencing Consortium and Celera Genomics each reported the first-draft sequences of the human genome. This announcement did not shake the world in the way it probably should have. It seemed to have escaped the attention of the majority of humanity that we now hold the key to every aspect of human physiology: how we look, to a great extent how we feel, and what our capabilities are. With this knowledge may come health and happiness—or mayhem—to many. Be this as it may, we cannot go back. We have taken a giant leap in the advancement of the human understanding of life. To quote Galileo, we have "seen what has remained hidden through the ages."

Forming the Code

Like the early scientist, you may find it incredible that only four organic bases can code for all of life. However, consider that computers store information as series of 0's and 1's. With only two options, all of the sophisticated computerized information manipulation is possible, including the graphic information in a computer game. It therefore shouldn't surprise you too much that all the biological information on the planet can be stored in only these four molecules:

- Adenine (A)
- Thymine (T)
- Guanine (G)
- Cytosine (C)

We should note that there is actually a fifth base involved, uracil. Thymine is found in DNA, and uracil replaces thymine in RNA.

The molecular structure of these molecules is given in Chapter 2. Very simplistically, the sequence of three of these molecules (called a *codon*) forms the code for one of the 20 specific amino acids that make up proteins. For example, the sequence GTT is the codon for glutamine. If you have a protein with 300 amino acids, the code will consist of 900 code molecules.

Again, the fact that the organic bases occur in pairs is fundamental to the information system. A piece of information can change hands many times without losing fidelity. Let's take our example of GTT. This sequence would be copied into another molecule as CAA (the matching pair for each base in the original molecule). If it is copied again from the CAA sequence, it will appear in its original sequence as GTT. (The T's will be U's if the molecule being formed is RNA).

See Table 3-1 for the codes. This, my friends, is the secret of life.

A strand of DNA consisting of AAA, ACC, CAA says "make a protein containing the sequence phenylalanine, tryptophan, and then valine." The same DNA sequence in any living organism will yield a protein containing the phenylalanine-tryptophan-valine sequence in the resulting protein. (This is actually quite a simplification, as you will learn in Chapter 7; however, it is true enough for now).

Overall, a given sequence of three of the nucleotides calls out the same amino acid for every living thing. (Of course, no statements regarding life are absolute... there are exceptions to this.) This is an extremely important point. This universality of the code allows you to insert human DNA into bacteria and have the code translated by the bacteria the same way it would be translated by a human. Also note that the code is robust—a given amino acid is usually specified by more than one code

Table 3-1 DNA Codes for Amino Acids

Amino Acid	mRNA Code
Alanine	CGA, CGG, CGT, CGC
Arginine	TCT, TCC, GCA, GCG, GCT, GCC
Asparagine	TTA, TTG
Aspartic acid	CTA, CTG
Cysteine	ACA, ACG
Glutamine	GTT, GTC
Glutamic acid	CTT, CTC
Glycine	CCA, CCG, CCT, CCC
Histidine	GTA, GTG
Isoleucine	TAA, TAG, TAT
Leucine	GAA, GAG, GAT, GAC, AAT, AAC
Lysine	TTT, TTC
Methionine	TAC
Phenylalanine	AAA, AAG
Proline	GGA, GGG, GGT, GGC
Serine	AGA, AGG, AGT, AGC
Threonine	TGA, TGG, TGT, TGC
Tryptophan	ACC
Tyrosine	ATA, ATG
Valine	CAA, CAG, CAT, CAC
Stop	ATT, ATC, ACT
Start	TAC

sequence. By specific arrangements of the four code words, all of the proteins known to the biological world can be called out.

The proteins carry out the work of the cell. They build structural molecules, control energy production, cause the making of secretary products, carry out the division of the cell, and control all of the chemical reactions that occur. So, if you can determine which proteins are made and in what quantities, you control the cell. Knowing the code, you can make whatever you want—just string together the amino acids you would like using the DNA code and build yourself a protein.

Coding for Proteins

Now you'll need a little more information about how all of this is accomplished. If you plan on building yourself a protein, it is essential that you understand how the old pro, the cell, translates the code on the DNA and makes a protein.

You need to understand the hardware for the code, which are the *nucleotides*, which are discussed in detail in Chapter 2. As you may recall, nucleotides are composed of two things: "structural" elements—a sugar (deoxyribose) and a phosphate—and an organic base (the information molecule). The sugar of one nucleotide hooks to the phosphate of the next nucleotide, forming a strong exterior scaffolding for the molecule. The information part is protected inside. One of your challenges as a biotechnologist is to figure out how to break up long strands of DNA.

To go from the code to a protein, at least the way a cell does, you need to do the following:

- Get inside of the very stable DNA structure (unzip it).
- Convert the genetic code into actual instructions to the cell.
- Have these instructions conveyed to an appropriate assembly point.
- Link these amino acids together according to the instructions.

Transcription

The process by which the genetic code is converted into instructions for the cell (the first two previous bullets) is called *transcription*. The process by which these instructions are read out to produce a protein is called *translation* (the third and fourth bullets).

To start the process you see in Figure 3-2, you have to get inside the DNA (unwind it and unzip it). The unzipping is done by an enzyme that is named *ribonucleic acid (RNA) polymerase*. (You will run into RNA polymerase again when we talk about common techniques in bioengineering.) RNA polymerase earns its name because it makes RNA (discussed later). The RNA polymerase unravels small pieces of the DNA, just involving the area where the code for a desired protein resides. In eukaryotes, the DNA is wound around proteins, *histones*, attached by hooks called *nucleosomes*. A complex combination of proteins remove the nucleosome, making the DNA accessible. Prokaryotic cells do not contain histones.

Once the DNA is unzipped, it can be read, that is, the genetic code can be decoded. Again, this process is called *transcription*. The reader is RNA polymerase.

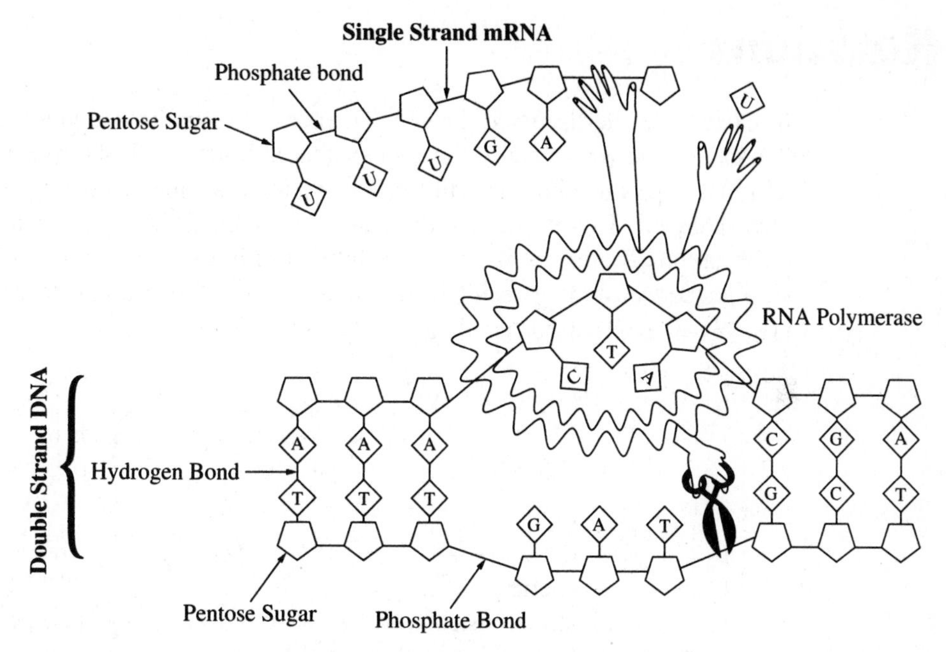

Figure 3-2 Transcription of DNA

This molecule travels down the DNA and builds up another molecule, RNA. RNA is very similar to DNA except it exists in much smaller molecules, the sugar is different (ribose instead of deoxyribose), and has uracil, another nucleic acid, in place of thymine. RNA usually exists as a single strand.

There are many types of RNA. The type that the RNA polymerase builds from the DNA code is called *messenger RNA*. mRNA is created because the sequence on the DNA results in changes within the RNA polymerase, such that binding sites are created for the specific nucleic acids that are complementary to the code on the DNA. These specific sites on the RNA polymerase molecule pick up the correct nucleic acid and string them in a sequence that mirrors the information on the DNA (except it holds the complementary nucleic acid for every DNA nucleic acid). The messenger RNA is the complement of DNA. Where DNA has cytosine, RNA will have guanine, etc. DNA pairs thymine with adenine but the RNA will pair uracil with adenine.

Consider the analogy of using a recipe (DNA) to prepare a dish (a protein). The RNA polymerase writes the recipe down and gives it to the messenger RNA to carry to the kitchen. mRNA is very important to you, bioengineer; it is carrying an important piece of the life code, the code for a specific protein. It is small and mobile, and it can move out of the nucleus.

Mutations

In Chapter 2, we discussed the consequences of errors in copying the DNA. Now you can see why it is that, if the code is altered or transcribed erroneously, the protein that is produced either will be dysfunctional or may have a different function than intended. This is because the code is no longer valid—it contains erroneous information. The error, like the one shown in Figure 3-3, may disrupt the start or stop sequence or may result in the substitution of one amino acid for another, possibly with disastrous results.

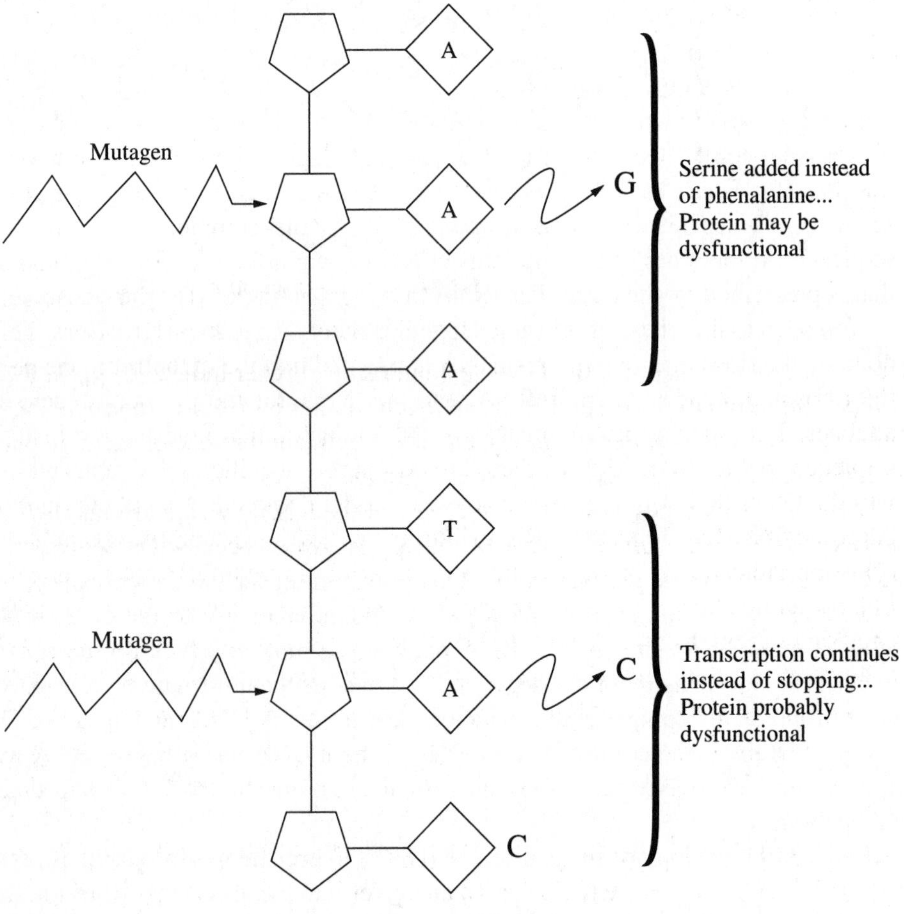

A
Mutagen
A
G
Serine added instead of phenalanine... Protein may be dysfunctional
A
T
Mutagen
A
C
Transcription continues instead of stopping... Protein probably dysfunctional
C

Figure 3-3 Mutations

Translation

The small mobile mRNA molecules convey the instructions for "cooking up" the protein to be conveyed to an appropriate assembly point. The assembly points are the *ribosomes*. These are compounds containing both RNA and proteins and exist in the cell cytoplasm. They act as a jig to hold the amino acids in place as the protein is being formed. In the analogy of a kitchen, the ribosomes are the cooks.

To begin the translation, the ribosomes attach to the messenger RNA. The resulting complex, the ribosomes and the mRNA, is relatively immobile. You need a mechanism to transport the assembly material (amino acids) to the assembly point. The transport is accomplished by another type of RNA, called *transfer RNA* or *tRNA*. The transfer RNA has been called the "Rosetta Stone of life." This molecule translates the nucleotide code on the mRNA into instructions for a specific amino acid. A given tRNA molecule is specific for a given amino acid. At one end of such a tRNA molecule is a binding site for a specific sequence of three bases on the mRNA. This site is called an *anticodon* because it is actually reading the mRNA sequence—a sequence, you remember, which is the complement of the DNA sequence. At the other end of the transfer RNA is a binding site for the amino acid that is prescribed by the code. The tRNA brings its amino acid to the ribosome.

You need to link these amino acids together according to the instructions. This is done by the ribosome. When the mRNA is released into the cytoplasm, one unit of the ribosome attaches to the mRNA, and the tRNA for the first amino acid also attaches. This specific tRNA must have the anticodon that matches the first base sequence on the RNA. That is, there are only a few specific tRNA molecules that will do. Other tRNA molecules are around, but their anticodon is not the right one. After the first tRNA joins the mRNA ribosome complex, a second, larger unit of the ribosome attaches and provides a holding niche for the next tRNA. The amino acids link by peptide bonds into a growing chain. The ribosome moves along the mRNA, chucking out the transfer RNAs that have already released their amino acid to the chain and attaching the appropriate tRNA into the mRNA. When not making proteins, the two subunits of the ribosome disconnect. See Figure 3-4.

Now you have successfully taken the recipe and produced the dish—a protein. As you can see in the kitchen analogy shown in Figure 3-5, you have used a storage site for the recipe, a way to read the recipe, a way to transport the recipe to the kitchen, a cook, and helpers to bring the ingredients to the cook.

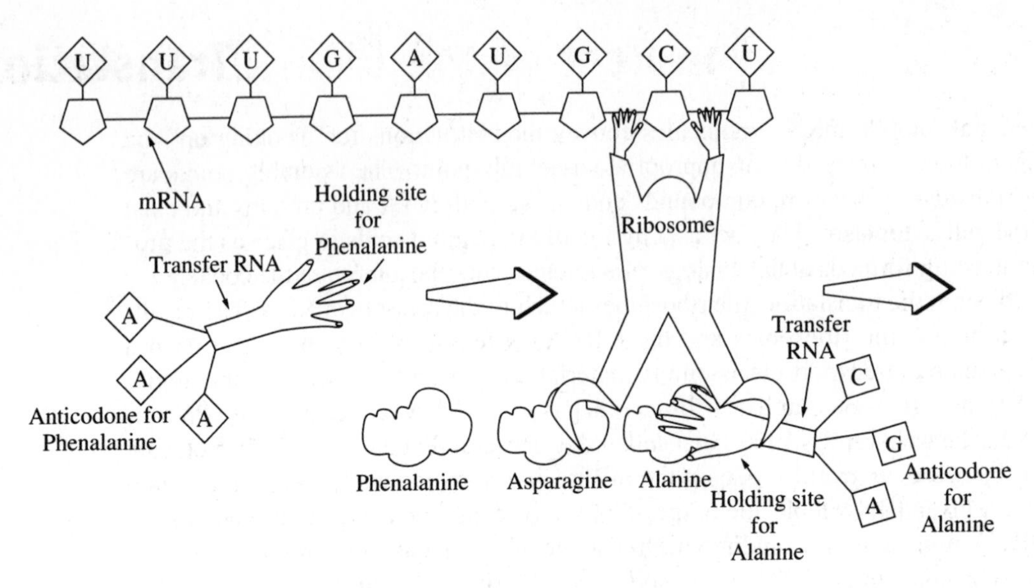

Figure 3-4 Translation of mRNA

Figure 3-5 Cooking up proteins

The Rest of the DNA...Is It Junk?

You may conclude that if you have the ingredients—DNA, RNA polymerase, nucleotides, ribosomes, and amino acids—you can make a protein. You may also realize that once you have the mRNA, you no longer need the DNA. You are partially correct, but there are some other control mechanisms you need to consider. First of all, most of the DNA does not contain genes! That's right, the overwhelming majority of the DNA, especially in a eukaryotic cell, is not "structural", that is, it doesn't code for anything. Some of it may be important just to hold the DNA in the right configuration for transcription. Or it may bind a repressor molecule that prevents transcription. Other portions of the DNA are sometimes treated as relevant information and sometimes are eliminated as irrelevant.

Multicellular eukaryotic cells have much more DNA than do little bacterial cells or single-celled eukaryotic organisms. This is one of the reasons the simpler cells are so much nicer to work with. A eukaryotic cell may have 1,000 times more DNA than a bacteria cell.

However, 90 to 95 percent of the DNA in the higher organisms is "noncoding"—it doesn't translate into anything. It usually consists of short stretches of random base pairs repeated many times. Other parts of the noncoding DNA form the hardy ends of the chromosomes, called *telomeres*, and also form the *centromere*, needed when the chromosome migrates into a new cell. Other noncoding DNA is interspersed among the coding DNA. Consider a given stretch of DNA that begins with a promoter site. This DNA will contain important instructional material (known as *exons*). However, it will also include non-instructional base pairs. Sequences found in the gene that are not present in the final mRNA product are called *introns*. So, when the RNA polymerase reads out a sequence of DNA, some of what is read out doesn't give a code that the transfer RNA will recognize as relevant. In fact, the intron portion of the gene segment tends to be much longer than the exon. Most genes of advanced organisms are split like this. After the messenger RNA is produced and before it is processed by the ribosomes, the excess portions must be removed or edited.

One Gene, One Protein...NOT!

Until recently, the concept of "one gene, one protein" was a mantra of genetics. The one-to-one association between the physical entity of a gene and the production of a protein was first described in 1902 by Archibald Garrod. He was a physician whose patients suffered from a non-life-threatening condition called alkaptonuria. These individuals produced urine that turned black upon exposure to air. The black color was due to an excess presence of a metabolic by-product homogentisic acid.

These patients lacked an enzyme to break down this metabolic by-product. Dr. Garrod noted that these people were frequently the offspring of closely related parents. The insights of Gregor Mendel on recessive versus dominant characteristics were well-known by this time. Dr. Garrod reasoned that the alkaptonuria disorder was due to a recessive gene and that this gene failed to produce the necessary enzyme.

"One gene, one protein" is, however, not entirely true. Certainly, each protein theoretically can be traced back to a gene. However, it turns out that a given gene may produce more than one protein. The final protein produced depends on the way that the mRNA is edited after transcription. The editing process may actually convert introns into exons, and vice versa. The phenomenon by which a given gene can produce more than one protein is called *alternative gene splicing*.

You may be asking yourself, why do it this way? Why not have a gene for every cellular protein? Well, it creates a complexity for us bioengineers. We would like to produce all proteins directly from DNA without worrying about ways the cell may thwart us by directing translation to produce an alternative protein. However, alternative gene splicing explains how the human genome that consists of an estimated 30,000 genes can code for many more different types of proteins than it has genes, approximately ten times as many.

The "decision" whether to excise or retain a portion of the mRNA probably depends on environmental factors. Consider lymphocytes that produce antibodies. Immature lymphocytes have not been exposed to the bacteria or whatever else it might be that its antibodies will attack (antigen). Antibodies produced by immature lymphocytes are attached to the cell surface. However, after the lymphocyte has been exposed to the target for its antibodies, it begins to excrete the antibodies outside of the cells. The cell first uses a form of the mRNA that causes the encoded molecule to be retained at the cell surface. In the presence of the antigen, the cell switches to an alternative splicing pathway, changes the protein slightly, and creates a protein that is secreted from the cell as a circulating antibody. Fascinating!

Control of Protein Production

The cell imposes elaborate controls to prevent production of proteins unless they are needed. This is because the production of a protein requires a substantial amount of energy.

Manipulating the ability of the RNA polymerase to bind to DNA is an important control on the production of proteins, as you can see in Figure 3-6. The initiation of the read-out of a DNA code for a protein is done in different ways. It can depend on the attachment of a molecule upstream from the gene. Such a molecule will cause the DNA to shift around physically to a configuration that allows the RNA polymerase

Figure 3-6 Control of protein production

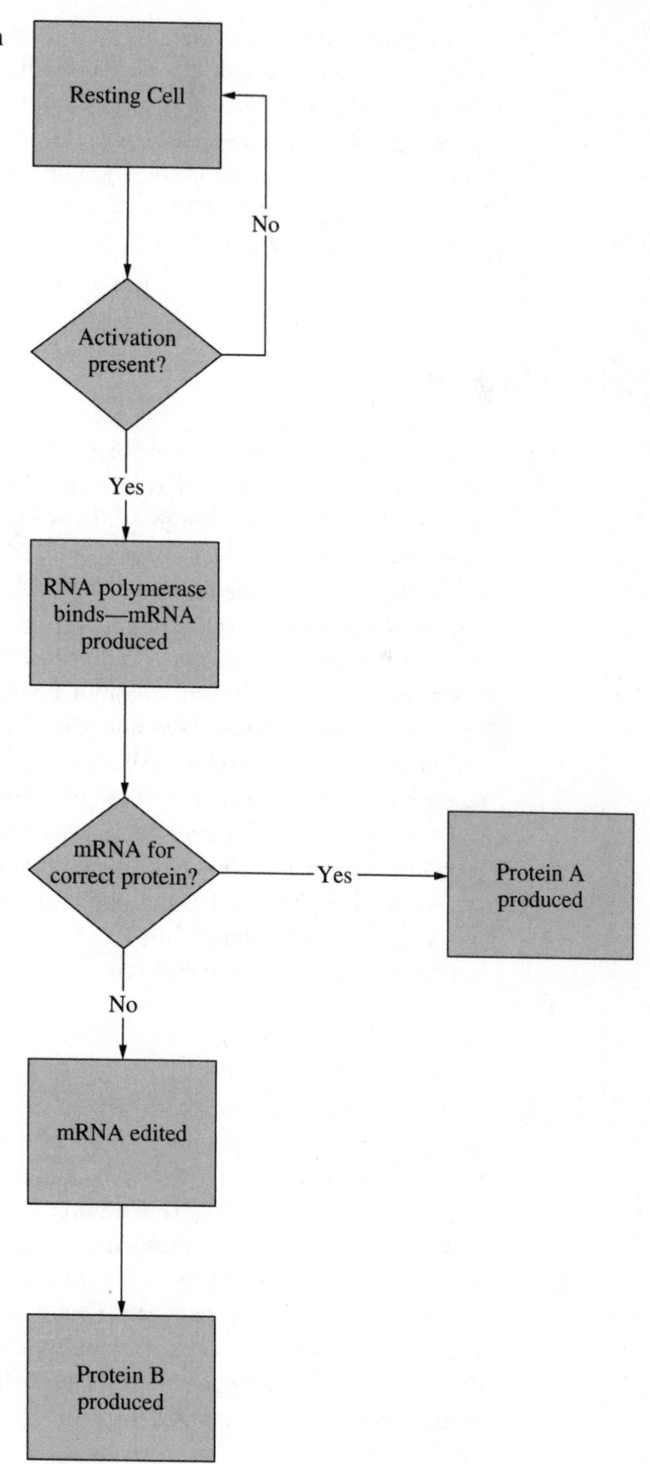

to attach. Alternatively, it can depend on the removal of a molecule, usually a protein, blocking the attachment site of the RNA polymerase. For example, liver cells are responsible for converting glucose into a form that is suitable for storage, called *glycogen*. The enzymes that do this conversion are "read out" from the DNA only when glucose levels are high. The elevated glucose levels cause a change in the physical configuration of the DNA. This change removes physical blocks on the promoter site so that the RNA polymerase can attach.

Summary

The foundation of life is a code made up of four organic bases called *nucleic acids*. The code is carried on DNA, a double-stranded structure that forms the familiar double helix. A series of three of these organic bases is the code for one of the 20 amino acids that form the structure of all proteins. The three-base unit that codes for an amino acid is called a *codon*. The series of these codons that codes for an entire protein is the *gene*. Genes have sequences at the beginning and at the end of the series that code for stop and start. In the midst of the genes is DNA information that is nonsensical and will not influence the final protein. This nonsense is known as *introns*, compared *exons*, which comprise the relevant information.

The transcription of the code on the DNA is done by a molecule called *RNA polymerase*. The RNA polymerase reads out the code by creating messenger RNA molecules. The RNA molecules are single-stranded nucleic acids that are composed of the same bases as the DNA with the exception of thymine, which becomes uracil in RNA. The RNA contains the complementary bases to the ones on the DNA, giving rise to the RNA. For example, if the DNA sequence says ATG, the RNA will say UAC. Because adenine always pairs with thymine (uracil in RNA) and cytosine pairs with guanine, you can read backwards from the mRNA to find out what the sequence is in the DNA.

Messenger RNA carries the code to ribosomes. The ribosomes recruit tRNA to obtain the needed amino acids. The tRNA contains an anticodon that pairs with the complementary bases on the mRNA. At the other end of the tRNA is a holding site for the appropriate amino acid. The ribosome captures the tRNA, one by one, as it marches down the mRNA, assembling a chain of amino acids as it goes.

An old tried and true of genetics, "one gene, one protein," has turned out to be not quite true. There are far more proteins produced by the human cell than there are genes in the human genome. This is possible because the cell edits some messenger RNA molecules after they are transcribed from the DNA. The protein that is ultimately produced depends on what portion of the mRNA is removed during the editing process.

Quiz

1. Translation:
   (a) is the process of translating the DNA code from the DNA.
   (b) is accomplished by RNA polymerase.
   (c) requires ribosomes.
   (d) requires removal of the nucleosomes.
   (e) a, b, and d are correct.
2. Ribosomes:
   (a) transport amino acids to the site of protein synthesis.
   (b) bind the anticodons of the transfer RNA.
   (c) act as a vice to hold the manufacturing apparatus together while the protein is being made.
   (d) transcribe the DNA code onto messenger RNA.
3. Proteins are encoded on the DNA by the following process:
   (a) each amino acid corresponds to a sequence of three organic bases on the DNA.
   (b) each protein corresponds to a sequence of ten organic bases on the DNA.
   (c) scientists are not sure how.
   (d) the code is done like a computer using a series of electrically-based gates.
4. Complementary base pairing:
   (a) varies with different species.
   (b) is important in producing mRNA.
   (c) is important in the duplication of DNA.
   (d) all are correct
   (e) b and c are correct.
5. The organic bases in nucleic acids:
   (a) are the same for RNA and DNA.
   (b) occur in random amounts in RNA and DNA.
   (c) are joined by hydrogen bonds on the interior of the double helix.
   (d) carry the genetic code.
   (e) c and d are correct.

6. A human gene:
    (a) codes for one protein.
    (b) consists of three organic bases.
    (c) depends on accurate translation by the RNA polymerase.
    (d) all are correct.
7. RNA polymerase:
    (a) makes mRNA.
    (b) must bind to the DNA.
    (c) assembles amino acids.
    (d) links amino acids together through peptide bonds.
    (e) a and b are correct.
    (f) c and d are correct.
8. The code for a protein:
    (a) is called a gene.
    (b) contains nonessential information embedded in the code.
    (c) must be read out by RNA polymerase to make a functional protein.
    (d) usually requires excision of unnecessary code after the code is read out.
    (d) all are correct.
9. Amino acids:
    (a) are usually coded by more than one sequence of bases.
    (b) are transported by messenger RNA to the DNA.
    (c) are assembled into proteins within the nucleus.
    (d) are held into place by the RNA polymerase.
    (e) all are correct.
10. The DNA:
    (a) must be partially unwound before the genetic code can be read out.
    (b) contains the same amount of adenine as thymine.
    (c) contains the same amount of cytosine as guanine.
    (d) one strand dictates the composition of the opposite strand.
    (e) all are correct.

CHAPTER 4

Genetics

Most people know that the way they look, and certain aspects of their health, are determined by their genes. However, most of us would be hard pressed to articulate the definition of a gene. You, bioengineer, are going to manipulate genes, so you need to understand that a gene is a DNA code for a specific protein. Within a species, you will find that same code for that same protein on the same spot on the same chromosome for every individual within that species. (Well, this isn't exactly true—we are genetically diverse and some genes can code for more than one protein, but it is true enough.) And as for the code, it is the same for all living things, algae to gorilla. A given sequence of organic bases on a DNA strand codes for the same molecule (or several molecules), no matter who you are.

A given gene occurs at predictable locations, called a *loci*, on the chromosome. There are different versions of a given gene within a population, and they are called *alleles*. For example, there are three alleles for blood types: A, B, or none (O), but there are only two spots (loci) where these alleles can occur, one on the chromosome you got from your mamma, and one on the chromosome you got from your papa.

Mendelian Genetics

Before we explain the basics of genetics, we need to give credit where credit is due. The basics were determined in the 19th century by a Gregorian monk named Mendel. Mendel didn't know what or where the code was, but he knew there was a code and he knew that the appearance of offspring was a combination of traits inherited from each parent. Mendel was aware that the inheritance of many characteristics, such as the height of an individual, was complex. However, to develop his system, he stuck to basic characteristics that he knew displayed simple inheritance patterns.

Mendel developed a model using pea flowers (see Figure 4-1), which are either purple or white. He observed that white-flowered peas that pollinated with white-flowered peas always produced seeds for white-flowered peas...hardly news. Also, purple-flowered peas that pollinated with purple-flowered peas always produced seeds for purple-flowered peas. The rub came when purple-flowered plants were pollinated with white-flowered plants and produced all purple-flowered plants. Of further mystery was the fact that the offspring from the latter (purple flowers from white and purple parents—all purple-flowered you remember), sometimes produced white flowers when pollinated with one another, proving that the seeds somehow carried a memory of a trait exhibited a generation or so ago.

Mendel developed a model for this, assuming that the traits were encoded in the cell on something called a gene. His model proved that some genes were stronger than others. The stronger genes were called *dominant*. The weaker genes were *recessive*. He had no idea of the mechanics of this, of course. In order to articulate his model, he assigned a letter to the trait under study. In the case of the color of peas, an appropriate letter would be p. So, the gene for purple was P and the gene for white was p.

If all ancestry has purple flowers, then any given individual carries the gene P from one parent and the gene P from the other parent. The articulation of the genes controlling a given trait is called the *genotype,* and in this case, it would be PP. The way the genotype is expressed—the appearance of the trait—is called the *phenotype*. Offspring from multiple generations of white flowers would have a genotype of pp.

If you mix the two, all the offspring have the genotype Pp. By convention, the first generation in such a study group is called the *F1* (this term has no relevance outside of a genetic experiment). Pp looks purple because the purple gene is dominant. Purple is the phenotype, or how it looks. If you pollinate Pp with Pp, you can get all possible combinations of P and p, like so:

	P	p
P	PP	Pp
P	Pp	PP

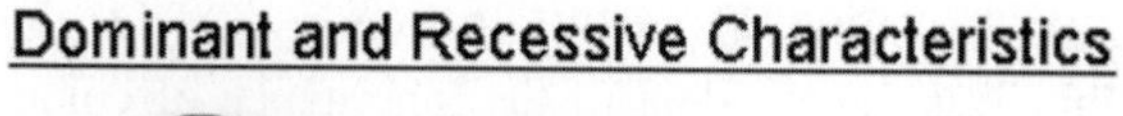

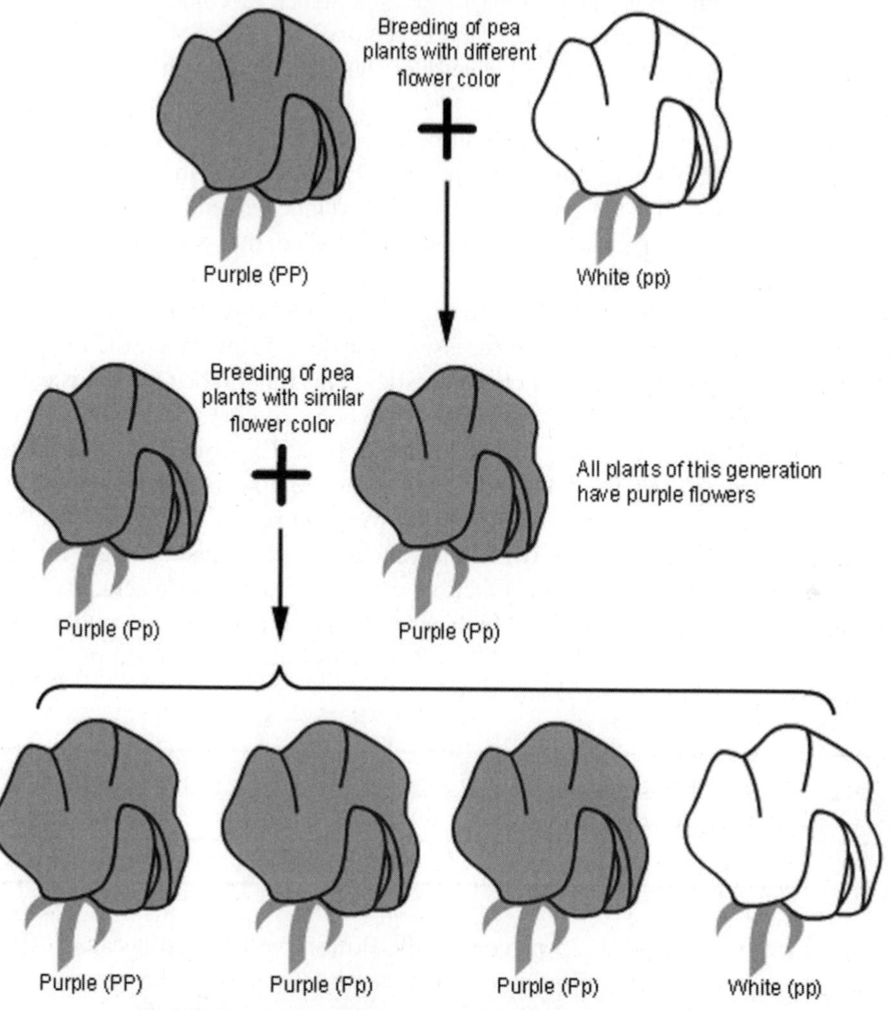

Figure 4-1 The simple model of purple and white pea flowers

This method of showing the distributions of the genotype of offspring is known as a *Punnett square*. Each one of the interior cells represents an offspring. You have one PP, two Pp's and one pp. All PP's and Pp's will be purple, so 75 percent of the offspring will be purple. Of course, there may be an F2, or any number of generations in the study, but things can quickly get complicated.

What about other pea traits? Consider height. Mendel had tall pea plants and dwarf pea plants. He looked at the height characteristic and found that it was inherited just like the color characteristic. Peas are either full sized or dwarf. He also found that the inheritance of height was independent of the inheritance of color. Remember that Mendel did not have the advantage of the 21st century knowledge of the mechanisms of this; for all he knew, the maternal genes were hooked together physically as were the paternal genes. Well, 21st century bioengineer, you know this is partially true, in that genes on the same chromosome will be inherited together. However, genes on separate chromosomes will sort independently. Either Mendel hedged his bet in studying traits that he had reason to expect would sort independently, or he was lucky. He was able to prove, at least for the traits that he studied, that genes were inherited independently of one another.

Even for traits that are controlled by a single gene, the expression of the genotype does not always follow a dominant/recessive model. In some cases, the offspring exhibits a melding of the parental characteristics. This inheritance pattern is called *incomplete dominance*. A classic example is the color of chrysanthemums. In the simplest model, the flowers may be red, white, or pink. Experimentation indicates that the red flowers carry the genome RR and the white flowers rr (just like the pea flowers). However, the genome Rr is pink, as you can see in the following table. Incomplete dominance is probably an unfortunate name for this because it implies that one gene tries to overcome the other but can't quite succeed. In actuality, most traits are a result of the expression of both genes. In the case of the chrysanthemums, one gene codes for a red pigment and the other codes for a white, or simply lack of pigment. When they mix, the result is pink.

	R	r
R	RR	Rr
r	Rr	rr

As shown in Figure 4-2, 25 percent of the flowers will be red because of the homozygous RR genotype. Fifty percent will be pink because of the heterozygous Rr genotype, and 25 percent will be white because of the homozygous rr genotype.

A familiar example of a trait that is inherited by the model of incomplete dominance is blood type. Remember that you have three alleles: A, B, and an allele that does not produce a surface antigen, O. Here are the possible genotypes:

	A	B	O
O	AO	BO	OO
A	AA	AB	AO
B	AB	BB	BO

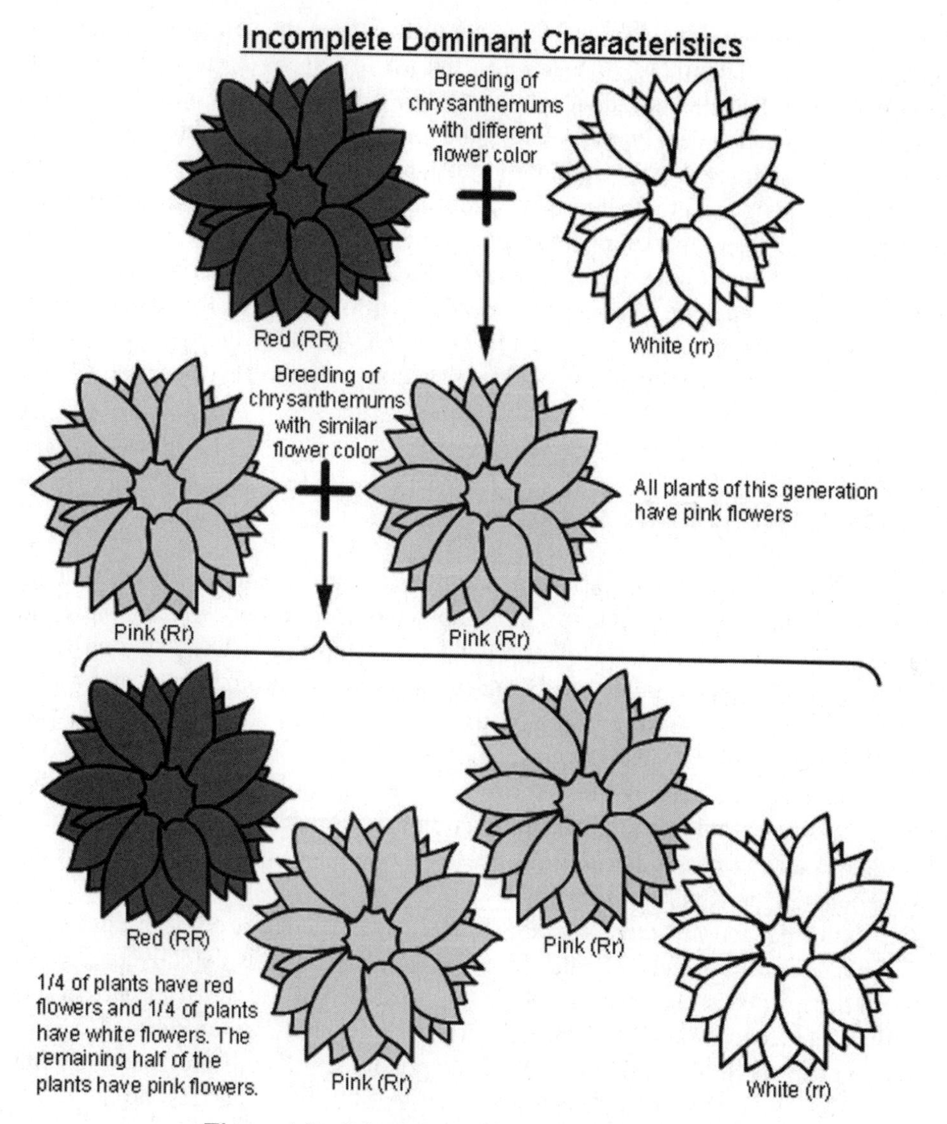

Figure 4-2 Inheritance of color in chrysanthemums

AO's and AA's will have blood type A. BO's and BB's will have blood type B. AB's will have blood type AB, and OO's will have blood type O.

Many healthy individuals are carrying recessive genes that will cause diseases in the homozygous condition (that is, if you have two of them). In many cases, normal levels of a protein require the same gene on both chromosomes—what we call the normal gene. I remind you that a gene codes for a specific protein. Some proteins affect things like the color of your eyes and others are essential for your health.

If either gene for the protein in question is defective, then the cell may not have the normal amount of the protein. An example is human Tay-Sachs disease. This is a very cruel condition that affects individuals of Middle Eastern Jewish ancestry. Babies with Tay-Sachs lack an enzyme (protein) called hexosaminidase A (hex A), which is necessary for breaking down certain fatty substances in brain and nerve cells. These substances build up and gradually destroy brain and nerve cells, until the entire central nervous system stops working. These unfortunate individuals have two recessive genes that normally code for the production of hex A. With two recessive, "defective" genes, none of the enzyme is produced. Individual with one "normal" and one "defective" gene (heterozygous) have reduced levels of hex A. The reduced enzyme levels are the basis of the screening test for individuals that carry the Tay-Sachs gene.

Sex-Linked Characteristics

This is a good segue into a discussion of *sex-linked* characteristics, or genes that are on the X chromosome. The X and Y chromosomes are the *sex chromosomes*, and other chromosomes are referred to as *autosomal* chromosomes. The Y chromosome carries very little genetic material. Females have two X chromosomes and males have just one. Therefore, any genes on the X chromosome in males are on their own. There are no paired genes to cover for them; there is only one copy. If the one gene is defective, defects will be expressed. The most common example is color blindness. Color blindness is much more common in males than females because the defect is due to a gene that causes cones in the retina of the eye to malfunction. This gene is on the X chromosome.

Let's consider a case where a woman is carrying a gene for color blindness on one of her X chromosomes. She may not be aware she has this defective gene. We will call her normal X chromosome X-C and the defective one X-c. Her partner has a normal X chromosome (X-C) and a Y chromosome, as shown in Figure 4-3.

As you can see in the following table, the genotype XX will produce a normal female. XX-C will produce a female who is normal but who carries the defective gene. XY produces a normal male, and X-CY produces a color-blind male. You can do a Punnett square to see the results of a cross between a normal female and a color-blind male (like the case shown in Figure 4-4). All the male offspring will be normal and all the female offspring will carry the gene for color blindness.

	X	Y
X	XX	XY
X-C	XX-C	X-CY

Color Blindness

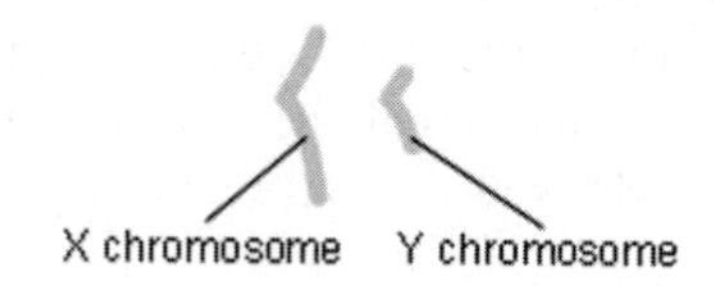

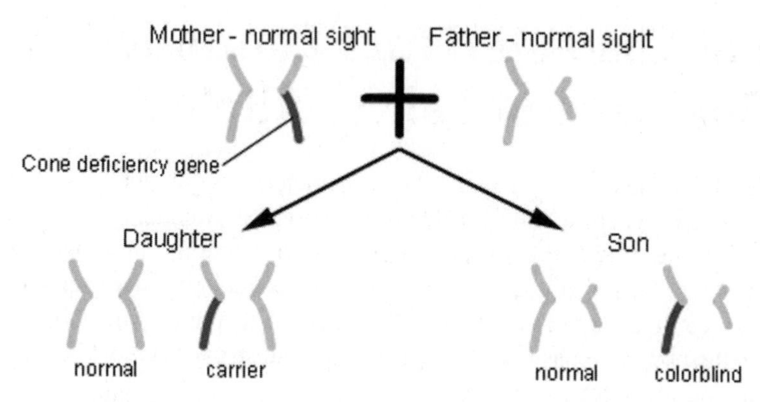

Figure 4-3 Inheritance of color blindness—carrier mother

Color Blindness

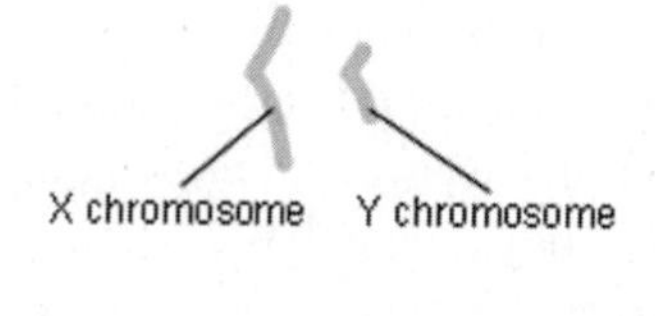

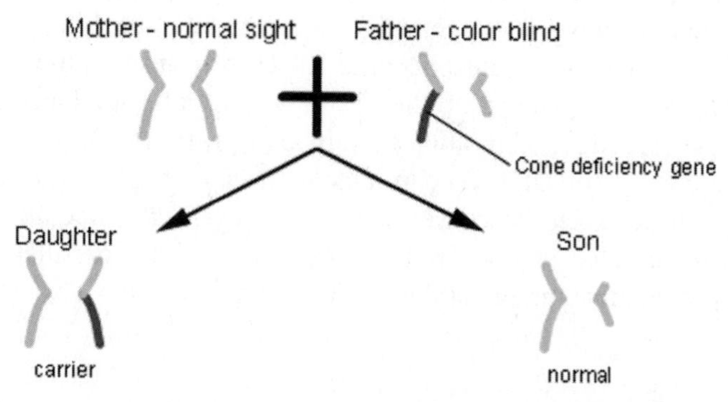

Figure 4-4 Inheritance of color blindness—color-blind father

Sex-linked characteristics are not to be confused with *sex-influenced* characteristics. There are many traits that are expressed differently in males than females even though the individuals have exactly the same genotype. The best example is male-patterned baldness (MPB). Males that are heterozygous for MPB will become bald at an early age, whereas females with the same genotype will retain a full head of hair until menopause. Height is also affected by the gender of the individual.

Inheritance of Characteristics Controlled by More Than One Gene or Pair of Alleles

I am sure that I don't need to tell you that most characteristics are much more complex than Mendel's tests. There was no overlap of phenotype in Mendel's studies, characteristics fit into one of two classes, and there was no blending in the heterozygote. On the other hand, Galton studied the inheritance of continuous characteristics, such as height and intelligence in humans, etc. He observed that any characteristic that is expressed in the population as a normal (bell-shaped) distribution must be controlled by more than one allele pair.

Consider the inheritance of human eye color. Eye color is determined by the amount of a pigment called *melanin* that is in the iris of the eye. Brown eyes have more pigment than blue eyes. The amount of pigment is determined by a number of genes controlling pigment production. Three gene pairs are known to control human eye color. Two of the gene pairs occur on chromosome pair 15 and one is found on chromosome pair 19. The bey 2 gene, found on chromosome 15, has a brown and a blue allele. The gene, located on chromosome 19 (the gey gene) has a blue and a green allele. A third gene, bey 1, located on chromosome 15, is a central brown eye-color gene.

Geneticists believe that the bey 2 gene has a brown and a blue allele. The brown allele is always dominant over the blue allele. The gey gene also has two alleles, one green and one blue. The green allele is dominant to the blue allele but is recessive to the brown allele on chromosome 15. This means that there is a dominance order among the two gene pairs. (See Table 4-1.) If a person has a brown allele on chromosome 15 and all other alleles are blue or green, the person will have brown eyes. If there is a green allele on chromosome 19 and the rest of the alleles are blue, eye color will be green. Blue eyes will occur only if all three alleles are for blue eyes. The inheritance of gray or hazel eyes and the factors that create multiple shades of brown, blue, green, and gray eyes are unknown.

Table 4-1 Inheritance of Eye Color

	Locatation			Color
	bey 1	bey 2	gey	
Allele	Blue	Blue	Blue	Blue
	Blue	Blue	Green	Green
	Brown	Blue	Blue	Brown
	Brown	Blue	Green	Brown
	Brown	Brown	Blue	Brown
	Brown	Brown	Green	Brown
	Blue	Brown	Blue	Brown
	Blue	Brown	Green	Brown

Genes clearly interact in very complex ways to produce human traits. So, biotechnologist, when the futurists among us imply that you will be able to construct the perfect human being based on your knowledge of the gene code, you are justified in rolling your eyes.

Mutations

The cell puts enormous resources into ensuring the integrity of its information systems; however, sometimes the code is corrupted. We call the code corruption *mutation*. To draw a computer analogy, a mutation inserts a virus into the code—a set of instructions that impairs the ability of the cell to function or even survive. Many of these "viruses," or mistakes, are eliminated immediately. By estimation, in the human organism, approximately a million mistakes (or mutations) occur every day. Clearly, there is a very efficient mechanism to eliminate defective cells. If there were not, the human genome would not last very many generations. However, occasionally, a bad code survives to propagate through the generations. Usually this is possible because the bad gene is present on only one chromosome, and the normal gene on the other chromosome can cover for the necessary function. Individuals with only one defective gene can live and reproduce. In this case, the gene is considered recessive and there must be two defective genes for the organism to be impaired.

How can the code become corrupted? Environmental factors can alter the composition of the DNA. A bolt of energy from radioactive emissions can knock out or alter a base of the DNA. Some chemicals can do the same, or they may alter a promoter or a terminator site. Chemicals that can induce mutations are called *mutagens*. Some (but not all) of these are linked with cancer; hence, they are carcinogenic.

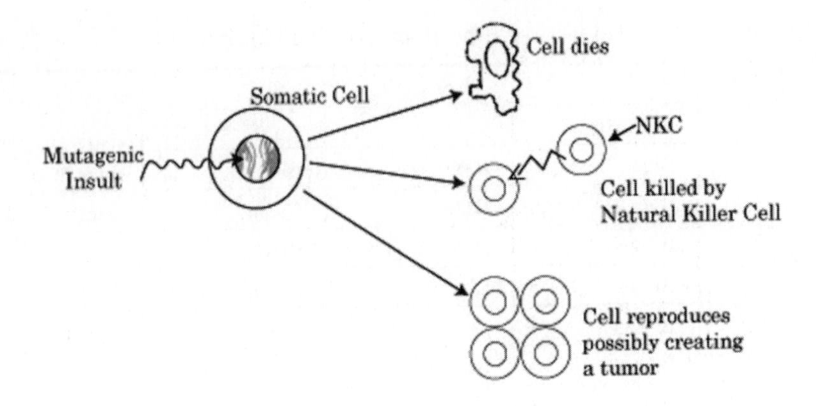

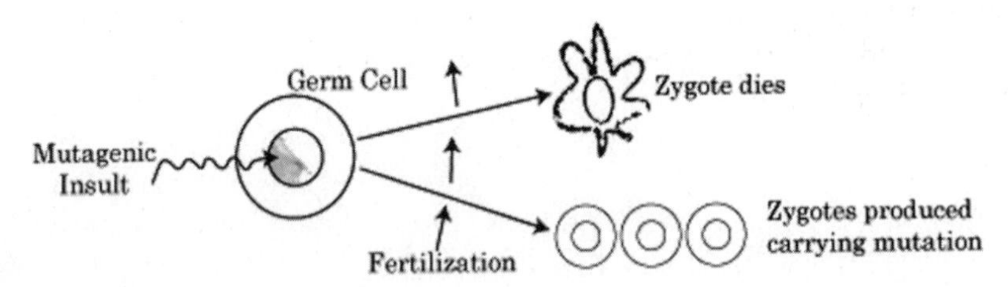

Figure 4-5 Mutations

If the alteration is in an autosomal cell (any cell other than those that produce sperm and egg), then the mutation may adversely affect the individual but has no way of passing to future generations. An example would be a mutation in a skin cell that results in the growth of skin cancer. The victim's offspring have no risk of inheriting skin cancer from this mutation. However, if the mutation occurs in progenitors of the germ cells, the sperm or egg, then the mutation may be transferred to future generations. The percentage of mutations that occur due to environmental factors versus simple mistakes in the code duplication process is controversial. Certainly, the cell is not 100 percent accurate in information duplication, and some mutations arise without external impetus. See Figure 4-5.

Genetic Diseases

How a disease is inherited was established for many conditions, well before the genes were identified, by observing how the condition was passed from generation

to generation. Many genetic diseases were observed intermittently in a family history and were therefore deduced to be recessive, according to Mendelian genetics. Remember that Mendel's experiments were conducted in the 19th century. The chances of inheriting a genetic condition are dependent on the race of the parents and part of the world they came from. You may be surprised at the number of individuals who suffer from such diseases.

CHROMOSOMAL DISORDERS

These type of disorders are less likely to show hereditary patterns because they are due to abnormalities in the chromosome structure. Most of these diseases involve the presence of an abnormal number of chromosomes within the cells. Cells may have abnormal chromosome numbers because of a defective cell division, either in the formation of the germ cell or early in the life of the zygote. One of the cells derived from a dividing cell receives an extra chromosome or two that should have migrated to the other cell. Disorders caused by excess chromosomes include Klinefelter's syndrome, which is due to one or more extra sex chromosomes (i.e., XXY); Down's syndrome, which is due to a *trisomy* (extra chromosome) of chromosome 21; and Trisomy 13 syndrome, which is due to an extra chromosome 13.

Klinefelter's syndrome is found in approximately 1 in 700 men. Klinefelter boys more often than other boys have delayed motor function, speech, and maturation development. Trisomy 13 syndrome occurs in approximately 1 in 10,000 live-born infants. Infants born with Trisomy 13 typically have a small head size (*microcephaly*), small eyes (*microphthalmia*), or sometimes absent eye or faulty development of the retina. About 80 percent of children with Trisomy 13 will have a congenital heart defect. Down's syndrome is the most frequent genetic cause of mild to moderate mental retardation and associated medical problems, and it occurs in one out of 800 live births, in all races and economic groups. Down's syndrome victims frequently exhibit hearing disorders, heart defects, and deficiencies in immune function. Most people with Down's syndrome have IQs that fall in the mild to moderate range of retardation.

Clinicians have had the ability to look at the chromosomes for over 40 years through a technique known as *karyotyping*. (See Figure 4-6.) In a non-dividing cell, the chromatin looks like a tangled mesh. However, during mitosis, the 23 pairs of human chromosomes are visible with a light microscope. To do a karyotype analysis, the cells are blocked in mitosis and stained with Giemsa dye. The dye stains regions of chromosomes that are rich in the base pairs adenine (A) and thymine (T) and produces dark bands. Each band may represent as many as 100,000 base pairs. Computer technology is used to digitally rearrange the chromosomes into pairs for analysis.

Karyotyping is used to screen fetuses for genetic diseases caused by chromosomal disorders. Cells representing the fetal genotype are collected from the amniotic

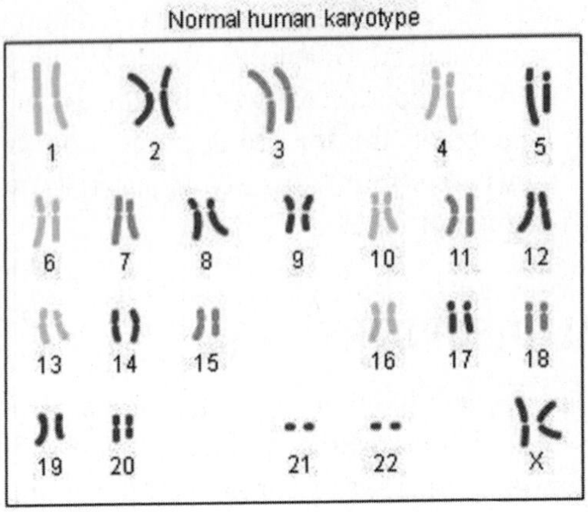

Figure 4-6 Normal human karyotype

fluid surrounding the fetus in the uterus, using a technique known as *amniocentesis*. The appearance of an extra chromosome is readily visible on the karyotype, as is shown in Figure 4-7 for Trisomy 13.

AUTOSOMAL RECESSIVE DISORDERS

These disorders are expressed only when the offspring receives a defective gene from each of the parents.

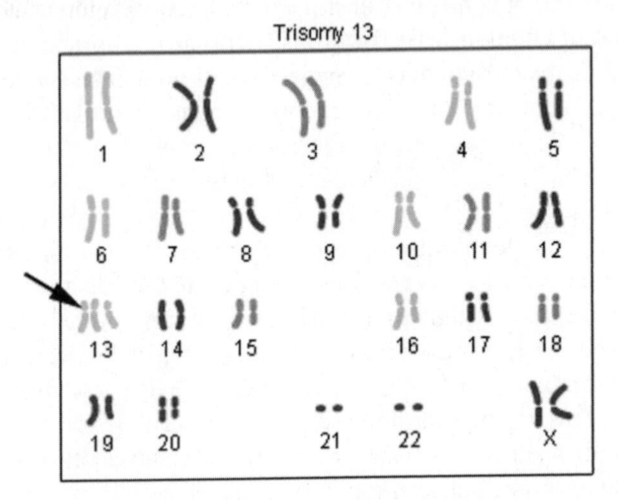

Figure 4-7 Karyotype of an individual with Trisomy 13

Cystic Fibrosis

Cystic fibrosis (CF) affects approximately 30,000 individuals within the U.S. It is the most common genetic disease among Caucasians, with one in 25–30 individuals carrying the gene and one in 3,200 live births expressing symptoms. The disease occurs because of the production of highly viscous mucus in the lungs, pancreas, and male reproductive tract. The result is lung disease, pancreatitis, and male infertility. Modern therapies have extended the life expectancy of CF victims from certain death in early childhood to survival to the early 30s.

Severe Combined Immunodeficiency Syndrome

Severe Combined Immunodeficiency Syndrome (SCIDS) is an extremely rare disease, resulting from the lack of a critical enzyme needed to metabolize excess adenosine. You have probably heard of children with SCIDS who live their lives in a germ-free bubble. They are isolated from their environment because without this critical enzyme, toxic products build up and attack the immune system. The victims of this disease have essentially no immune function. Fortunately, these sufferers no longer live their lives in an isolated "bubble" but are provided with regular supplements of the missing enzyme coated on beads.

Phenylketonuria

Phenylketonuria is a fairly common condition (1 in 12,000 Caucasians) and results from an inability to metabolize the amino acid phenylalanine. All babies in the U.S. are tested for high levels of phenylalanine and, if they test positive, phenylalanine is withheld from their diet. Diet restrictions allow these individuals to lead a normal life. Otherwise, they develop severe mental retardation within months after birth.

Sickle Cell Anemia

Sickle cell anemia occurs primarily in individuals of African descent. It occurs from a defect in red blood cells that causes them to lose the flexible disc shape of normal cells and take the form of a rigid sickle. These cells are unable to pass through capillaries. For individuals with the defect in both genes, the sickle cell anemia gene is definitely a bad thing—they will not survive to adulthood without intervention. Individuals with only one defective gene experience the defect in only some of their red blood cells and can survive the disease. Interestingly, individuals with only one bad gene (heterozygous) have an increased resistance to malaria. It seems that the organism that causes malaria lives in red blood cells and finds the sickled red blood cells to be an inhospitable environment. The phenomenon whereby the heterozygous condition actually enhances survival of the population is of great interest to a population geneticist. Not surprisingly, carriers of the sickle cell anemia gene tend to derive from areas with high incidence of malaria. This documented positive effect of a heterozygous genotype for a mutation gives us biotechnologists pause for thought: if genes that we perceive as "bad" are eliminated from the human gene pool, we may affect other functions in ways we don't anticipate.

AUTOSOMAL DOMINANT DISORDERS

The gloomy list just presented gives examples only of genes that are fatal to their carriers if inherited from both parents, i.e., are recessive. There are other genetic diseases that may be inherited from only one parent—those considered *autosomal dominant* and those carried on the X chromosome. The dominant conditions not on the X chromosome include thalassemia and Huntington's disease.

Thalassemia

The hemoglobin of victims of thalassemia is defective. (See Figure 4-8.) This disease is expressed in every individual that carries the defective gene because formation of normal hemoglobin requires input from both chromosomes. Hemoglobin is formed of four large polypeptide molecules, two of which are called *alpha globulin* and *two beta globulin*, each exquisitely conformed to hold the iron atom just right for oxidation and reduction under appropriate circumstances. The thalassemia gene is involved in the formation of two of these molecules, the beta globulins. The information to construct the beta globulin molecule comes from both chromosomes; half of the molecules are formed according to the instructions on one of the chromosomes and half on the other. With thalassemia, half of the beta globulins in each of two positions in the hemoglobin molecule are defective. This means that the chance of having a normal hemoglobin is only 25 percent (50 percent chance in one position *times* 50 percent chance in the other position *equals* combined probability of 25 percent). The blood's ability to deliver oxygen is severely compromised.

This condition serves to make an important point. In some cases, the normal function of one chromosome is all that is necessary to have a fully functioning cell. You can see how this might be accomplished if you had a very effective feedback mechanism. Let's consider induction of the enzyme lactose dehydrogenase in bacteria. This enzyme is necessary for metabolism of the sugar lactose, found in milk. The presence of lactose releases the gene for lactose dehydrogenase for transcription. I say "releases" because the gene is usually blocked and cannot be translated to produce a protein. However, if lactose is present, the lactose binds to the inhibiting protein that is fixed on the chromosome and is blocking gene readout. Who needs

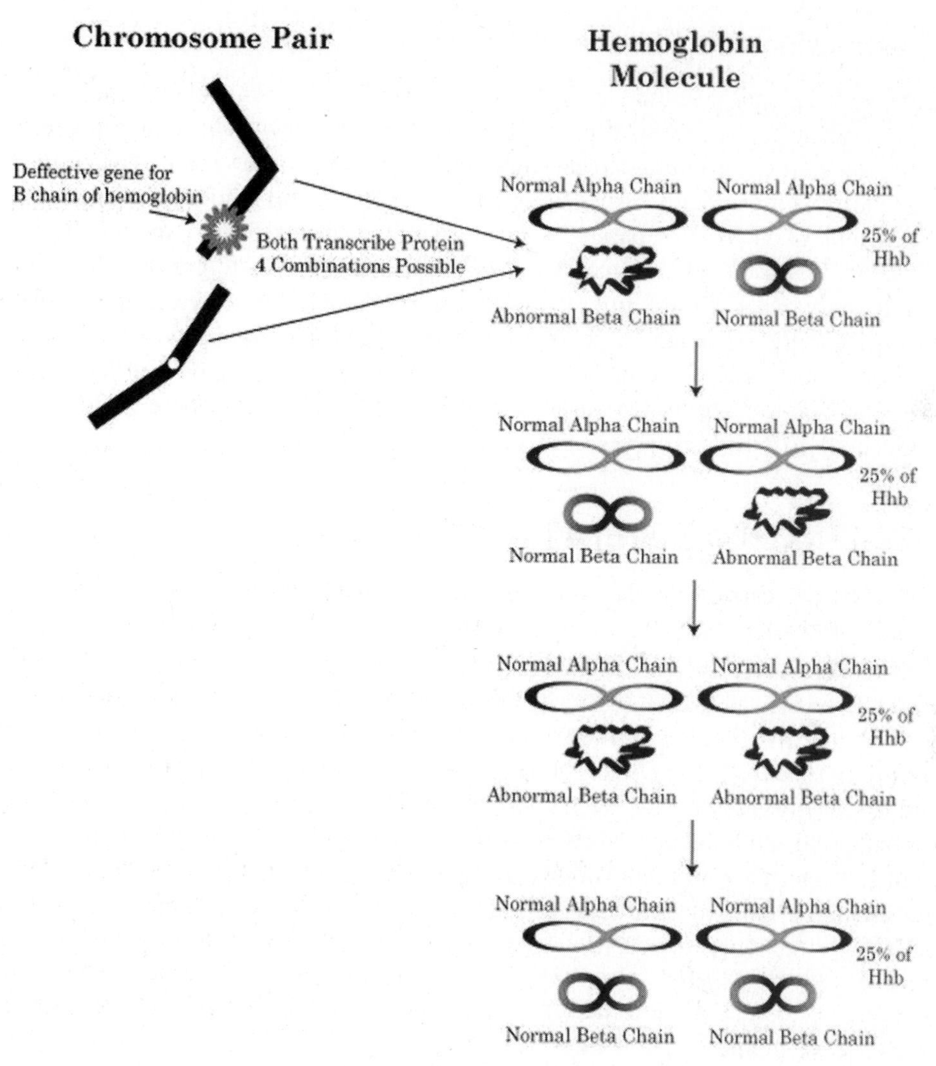

Figure 4-8 Thalassemia

lactose dehydrogenase if there is no lactose? However, when the inhibiting protein is tied up with the lactose molecule rather than sitting on the chromosome, the lactose dehydrogenase can be produced. As soon as the lactose levels fall, the inhibiting protein returns to its binding site on the chromosome. If you have a defective lactose dehydrogenase gene on one chromosome, the other chromosome will still be induced to produce the enzyme until the lactose level falls; it just might take a little longer. However, in cases where the gene function is more complex, you might need both chromosomes. In the case of hemoglobin, production is more or less nonstop, and both chromosomes need to work full time.

Huntington's Disease

Although the gene resulting in Huntington's disease has been isolated, the way by which this gene causes this dreadful condition is not known. The symptoms result from genetically programmed degeneration of neurons in the brain. The result is uncontrolled movements, loss of intellectual faculties, and emotional disturbance. Each child of an afflicted parent has a 50 percent chance of developing the disease. Since the disease does not manifest itself before the third or fourth decade of life, many individuals with Huntington's disease have reproduced before they realize that they carry the defective gene. Modern screening techniques will identify the defective gene for individuals that submit to screening. It is on the fourth chromosome and consists of 40 to over 100 repeated CAG segments, where normal individuals have 26 or less copies of this trinucleotide.

SEX-LINKED DISORDERS

The diseases carried on the X chromosome include color blindness (more of a condition than a disease) and leukodystrophy. Leukodystrophy is a cruel disease that causes neurological degeneration, usually striking young boys. When the term "dystrophy" is used in medicine, it is meant to imply a condition that is genetically determined and that is progressive; that is, the condition tends to get worse as the patient gets older. Leukodystrophy results in progressive abnormalities in the myelin sheath of the nerves in the brain. The pattern of inheritance in leukodystrophy is either autosomal recessive or X-linked. In autosomal recessive disorders, boys and girls are affected equally, and both parents must be carriers (heterozygotes). Carriers have no disability. When two carriers marry, on the average, one half of their children will also be carriers, one quarter will have the illness, and one quarter will be entirely normal. The sex-linked conditions exhibit the typical pattern of expression in half of the male offspring. The female heterozygous carriers sometimes develop mild disabilities as well.

Influence of Environment

The previous examples would lead one to conclude that the genome leads to inevitable outcomes regarding the individual's health. In actuality, environmental factors and complexities in gene regulation introduce uncertainties in the prediction of gene expression. Even in conditions with dreadful prognosis such as Huntington's disease, there is individual variability in the age of onset and progression of the disease.

Gene Therapy

"Gene therapy," meaning inserting the missing piece of code (gene) into the patient's own cells, is the source of great hope for suffrers of these diseases. The appropriateness of gene therapy assumes that (1) the etiology of the disease is clear (i.e., we know what causes it) and a substantial contributor to the cause is a gene, and (2) we know which gene is the problem. It is best to clearly understand how the particular misinformation is coded on the aberrant gene and how this coding results in the symptoms exhibited by the patient. Consider the case of phenylketonuria. The genetic cause of this disease is clear, and inserting the missing genetic instructions to produce the missing enzyme would be a reasonable course of action.

By contrast, consider the inheritance of diabetes. Type I diabetes is clearly a genetic disease. The risk of diabetes increases by a factor of 10 for people with a parent, sibling, or child with diabetes. If you have a diabetic identical twin, your chance of developing diabetes is one in two. However, a gene for diabetes has not been found. The development of the disease has been associated with the presence of a type of gene called HLA (human leukocyte antigen) genes. Individuals with two of the suspect forms of the HLA genes are more likely to develop diabetes. But the development of diabetes is not inevitable, and some diabetics apparently don't have these genes. Clearly, we are in no position to provide gene therapy to diabetes suffrers.

Many clinical trials are underway to develop gene therapy strategies on diseases that are clearly due to a defective gene. Current strategies in many cases use a virus vector as a means to insert the missing gene—usually a gene that codes for the correct protein—into the patient's cells (see Chapter 5). This approach has met with considerable success in clinical trials involving SCIDS patients. However, other clinical trials are reporting varying success. For example, Targeted Genetics Corporation recently announced that it was ending a cystic fibrosis gene therapy study because of lack of significant clinical improvement on the part of their subjects. Also, when two of the individuals treated for SCIDS later developed leukemia, the scientific community raised concerns about the uncertainties of randomly inserting genes into people.

Even though the revolution in health care through gene therapy has yet to arrive, the future will probably see increased use of this strategy. Even the failed trials are adding to our knowledge and improving chances for successes in the future. However, clearly we have a lot to learn about genetic manipulation of human patients.

Summary

Basic patterns of inheritance were defined in the 19th century by Gregory Mendel. Mendel described genes that were "hidden" in an individual's genome, as recessive genes. Because there are two genes for every trait, one from each parent, some characteristics can appear normal if only one of the genes is normal and the other gene is defective. However, if an individual inherits two defective genes, this individual will exhibit the abnormal characteristic. Such an inheritance pattern is called *dominant/recessive*. The recessive characteristic doesn't necessarily have to be abnormal. In Mendel's famous pea experiments, the recessive trait was for white, as compared to purple, pea flowers. Blue eyes are recessive to brown but are not abnormal. Sometimes the trait you don't want is dominant to the normal trait. Other characteristics will exhibit an in-between state. This pattern of inheritance is called *incomplete dominance*. Mendel's example was chrysanthemums; red chrysanthemums bred with white chrysanthemums will produce pink flowers.

Many traits are controlled by multiple genes, such as human height. Any characteristic that is found in a bell-shaped distribution in the population can be assumed to be controlled by multiple genes. An example is intelligence. Clearly, understanding the genetic control of characteristics dictated by multiple genes is much more complex than the Mendelian traits. We are still unclear about what exactly determines human eye color, for example.

Characteristics carried on the X chromosome are known as *sex-linked*. Females have two X chromosomes and males have only one. Therefore, recessive characteristics carried on the X chromosome are much more likely to be expressed in males than females. A classic example is color blindness.

Many disease conditions arise because of mutations in the gene code. Mutations can be caused by energetic interruptions, such as ionizing radiation, or by chemicals. Not infrequently, the genetic replication mechanism makes mistakes, leading to mutations. Cellular mechanisms exist to eliminate defective or mutated cells, although obviously some survive. The only mutations that can be passed to subsequent generations are mutations in the germ cells, cells that lead to sperm or egg.

A discouraging number of adverse human conditions are linked to genetic defects. Some of these are due to abnormal separation of chromosomes—Klinefelter's syndrome, Trisomy 13, and Down's syndrome. Human genetic diseases due to recessive genes include cystic fibrosis, phenylketonuria, sickle cell anemia, and Severe Combined Immunodeficiency Syndrome (SCIDS). Human genetic diseases due to dominant genes include thalassemia and Huntington's disease.

Gene therapy is possible with the development of recombinant DNA technology. This is applicable only when the genetic mechanism for the disease is well understood. Theoretically, the normal gene, missing in individuals with recessive genetic disease, can be inserted into the patient's own cells. This approach has already met

with success in treating infants with SCIDS. However, there are many technical obstacles to overcome, and other clinical trials have not been as promising. Nonetheless, gene therapy may thwart nature in some of her cruelest tricks, and it provides hope to many a desperate parent.

Quiz

1. Alleles are:
   (a) different forms of a given gene within a population.
   (b) the different appearances (phenotypes) due to a given gene.
   (c) the defective forms of a gene.
   (d) different places on a chromosome where a gene may appear.
2. Sex-linked conditions are manifested more frequently in males because:
   (a) the males are more likely to have two of the defective genes.
   (b) males have only one X chromosome and therefore only one copy of the gene in question.
   (c) females are protected by their hormonal structure.
   (d) only males can inherit the defective chromosome from their mothers.
3. Sickle cell anemia:
   (a) is an example of incomplete dominance.
   (b) is an autosomal recessive disease.
   (c) causes impairment in both homozygous and heterozygous conditions.
   (d) b and c are correct.
4. An individual with blood type A:
   (a) may carry the gene for B.
   (b) may carry the gene for O.
   (c) must have two A genes because A is recessive.
   (d) a and b are correct.
5. Autosomal dominant diseases:
   (a) are found on the X chromosome.
   (b) are not influenced by environmental factors.
   (c) are expressed in all bearers of the defective gene.
   (d) are not candidates for gene therapy.

6. Type I diabetes:
   (a) is inherited as an autosomal recessive disease.
   (b) is not a genetic disease.
   (c) is amenable to gene therapy.
   (d) is clearly at least influenced by genetic factors.
   (e) a and c are correct.
7. Human traits controlled by multiple genes:
   (a) will be more difficult to manipulate through genetic engineering.
   (b) typically exhibit a bell-shaped or normal distribution in the population.
   (c) are rare.
   (d) a and c are correct.
   (e) a and b are correct.
8. Mutations:
   (a) are not found in a natural state.
   (b) occur frequently in a normal organism.
   (c) can result in cancer.
   (d) may result in a cell that is not viable or that is destroyed by the immune system.
   (e) c and d are correct.
   (f) b, c, and d are correct.
9. If you are a color-blind male:
   (a) all of your sons will be color-blind.
   (b) none of your sons will be color-blind.
   (c) all of your daughters will be color-blind.
   (d) a and c are correct.
10. If you are a male and have sickle cell anemia:
    (a) all of your sons will have sickle cell anemia.
    (b) assuming your mate is normal, half of your children will be carriers.
    (c) all of your daughters will be carriers.
    (d) assuming your mate is normal, 25 percent of your offspring will have sickle cell anemia.

CHAPTER 5

Immunology

Think of the immune system as an army of many types of cells deployed throughout the body to mount a defense against anything recognized as foreign. Its soldiers exist in the blood, in the lymph tissues, and throughout the body organs. The immune systems is exquisitely designed to eliminate foreign matter, either by recognizing the material specifically as foreign and grabbing it for disposal or by making the environment deathly unpleasant. The immune system also monitors cells produced within the body to ensure that they are healthy. Our bodies will eliminate abnormal or infected cells.

The immune system originates in the bone marrow, where progenitors of most of the immune cells are produced. These include all the cells commonly referred to as white blood cells or *leukocytes*. White blood cells include *lymphocytes* that produce antibodies, *monocytes* that become *macrophages* in the tissues and *neutrophils* that secrete chemicals unpleasant to anything nearby. If lymphocytes mature in the bone marrow, they are known as *b-lymphocytes*. If the cells migrate to the thymus to mature, they become what are known as *t-lymphocytes* (*t* for thymus). Only the B-cells produce antibodies. T-cells are a kind of broker in the immune response, as discussed next.

Many of these cells move between the blood and the *lymphatic system*. The lymphatic system parallels the vascular system and collects the fluid that accumulates

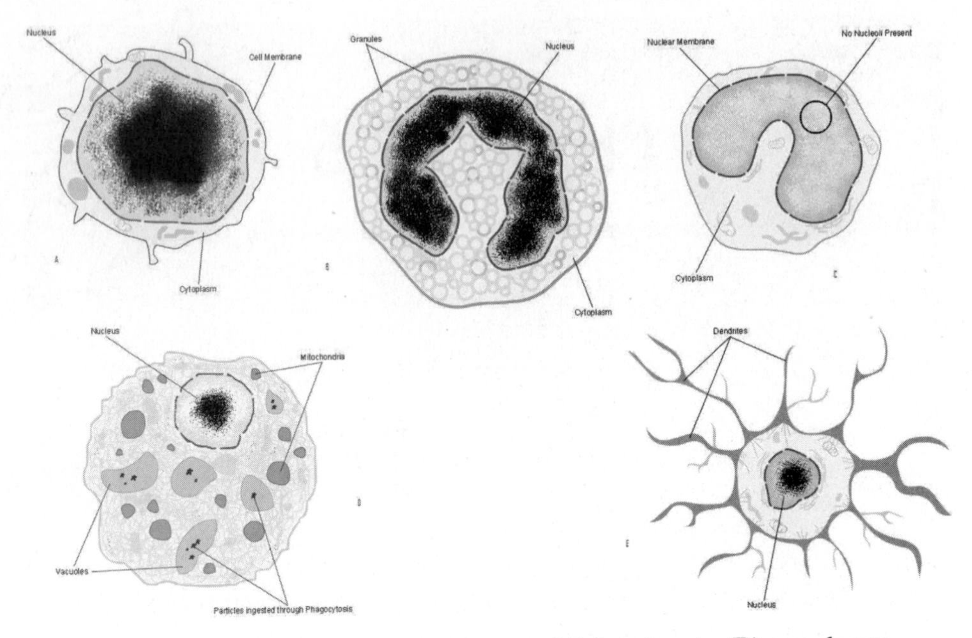

FIGURE 5-1: Cells of the immune system: (A) lymphocyte, (B) granulocyte, (C) monocyte, (D) macrophage, (E) dendritic cell

between cells (interstitial fluid). The thin tubes of the lymphatic system are interrupted by the *lymph nodes* found throughout the body. These nodes act as a filter and swell in the presence of an infection. The "naïve" lymphocytes take up residence in these nodes. They are considered naïve if they have never encountered the antigen for which they were designed and against which they can produce antibodies.

The basic cellular components of the immune system are depicted in Figure 5-1:

- **B-lymphocytes** Antibody-producing cells. Each B-lymphocyte can produce only one type of antibody.
- **Dendritic cells** Antigen-presenting cells derived from either lymphocytes or monocytes. These are professionals at capturing the target of interest and showing it in just the right way to activate other cells.
- **Macrophages** Large amoeboid cells that ingest and destroy foreign material. They start out life in the bone marrow and blood as monocytes.
- **Monocytes** Mononuclear cells that are found in the blood stream and mature in the tissues as macrophages or dendritic cells.
- **Natural killer cells (NKCs)** These cells attack the membranes of defective or infected cells, resulting in their destruction. Unlike cytotoxic T-lymphocytes, these do not require recognition of specific antigens on the cell surface.

- **T-lymphocytes** Cousins of the B-lymphocytes. These cells don't produce antibodies but they are essential for antibody production and for activation of certain types of macrophages.
    - **Cytotoxic T-cells** Part of the "cell-mediated response," they kill abnormal or infected cells.
    - **Inducer T-cells** Mediate the development of T cells in the thymus.
    - **Suppressor T-cells** Help control the magnitude of the response.
    - **Helper T-cells** Important in initiating and maintaining the B-lymphocyte response to an antigen.

Antigens

Antigens are any molecules of such a form that they can initiate an immune response. Such molecules are said to be *antigenic*. Antigens may be proteins, lipids, or carbohydrates. Note that each antigen may have a number of distinct sites that are individually recognized by the immune system, known as *epitopes*, which are the smallest units of the immune system.

Remember that cells have numerous receptors sticking out of their surface. The diverse molecules on the surface of cells can be antigenic. Bacteria have numerous surface molecules that your immune system will recognize as foreign. Even though your own cells also have surface antigens, your body has learned not to "see" these antigens. If you receive cells from someone else, as in a blood transfusion or an organ transplant, your immune system will respond to the antigens that are new—the ones you do not share with the donor. The immune system will attempt to reject the new tissue. Antibodies are produced that will lock onto specific antigens on the tissue. The resulting reaction will cause the new tissue to die.

The immune system learns early in the development of the organisms to distinguish between foreign antigens and the antigens of its own body, self-antigens. If the immune system attacks its own, the result is an *autoimmune disease*. To attack, in general, means to produce antibodies against the antigen. (See Figure 5-2.) For example, multiple sclerosis is a degenerative disease caused when the individual's own immune system attacks self-antigens within the nervous system. The victim experiences a gradual loss of motor skills as the nerve cells are killed by their own immune system.

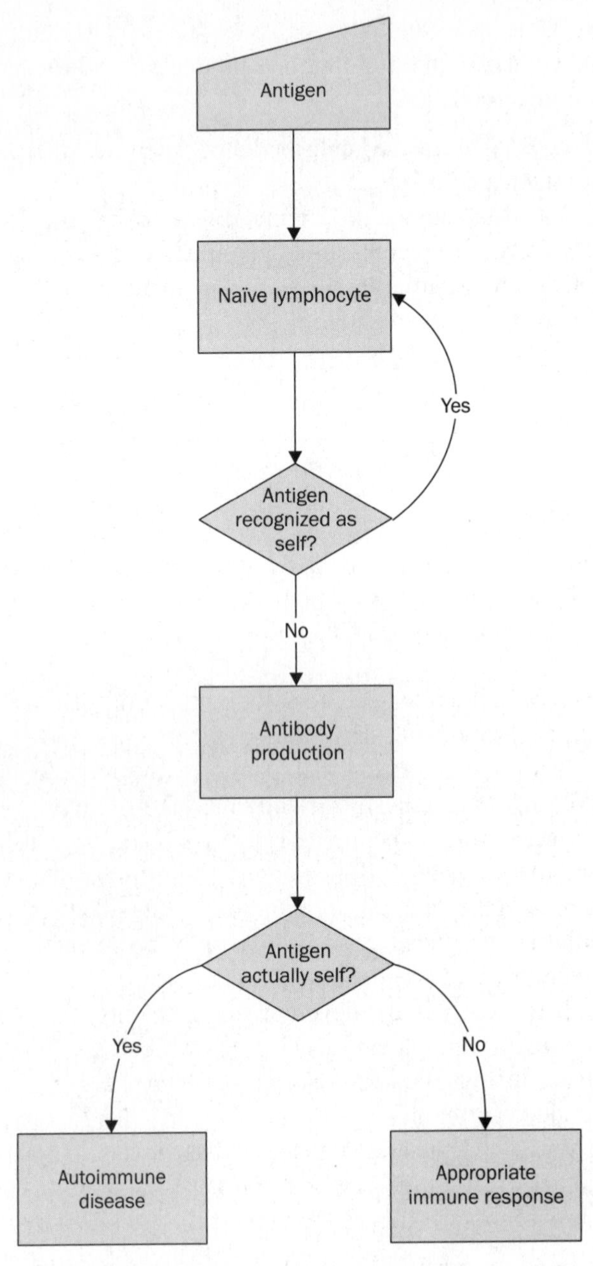

FIGURE 5-2: Development of autoimmune disease

Antibodies

Antibodies are molecules produced by B-lymphocytes. Their sole purpose is to adhere to material that is foreign to the body. Each antibody has affinity to a molecule or group of molecules forming a very specific configuration. In this way, antibodies differ from *antibiotics* that your doctor gives you to fight infection. Antibiotics are generally toxic to a class of organisms. For example, penicillin is toxic to all gram-negative bacteria. An antibody would be specific to a certain site or antigen on the surface of only one species of the bacteria.

Antibodies were called globulins when they were first isolated and then *immunoglobulins* when they were associated with immune function. Further research identified the configuration associated with most antibodies in the blood as "gamma" configuration. The general name for antibodies then evolved into *gamma globulins*.

Each B-lymphocyte can produce only one specific antibody. Until the B-lymphocyte encounters its antigen, it is in a resting state (called naïve) with its antibodies sticking hopefully from its surface. When an antigen encounters a B-lymphocyte with a "matching" antibody on its surface, the antigen is bound is bound to the antibody. The union of antibody and antigen has been likened to lock and key, or hand and glove. Once the antibody/antigen complex has been formed, the B-lymphocyte is activated. (See Figure 5-3.) The antibodies it produces are modified such that they no longer simply protrude from the cell surface but are cranked out into the blood and lymph system. The B-lymphocyte begins to vigorously divide, forming colonies, called *clones*, of identical cells, all producing the same antibody. When the antigen has been eliminated, the B-lymphocyte colonies quiet down, but the population of this specific antibody-producing lymphocyte remains enlarged. If the antigen is encountered again, then an amplified response occurs. The ability to mount the amplified response is the source of *immunity* to a given bacteria.

You may wonder, given that antigens are foreign, how the body designs specific receptors for a potentially infinite assortment. Actually, the process is random. The foreign protein just happens to fit one of the hundreds of thousands of receptors that we are all born with and that are seeded throughout the immune system. Certainly, the chance possession of some of these could have an evolutionary advantage. Consider a population that is continuously exposed to smallpox. Through the generations, such a population might build up a reservoir of lymphocytes with smallpox antigen receptors. The population would evolve a resistance to smallpox.

B-lymphocytes undergo a very unique process as they mature. Their genetic material actually shifts around. The DNA sequences that determine the structure of the antigen recognition and binding regions of the antibody structure are shuffled. This molecule mix-and-match is called *somatic rearrangement*. You are not stuck with what your parents give you, immunologically speaking. The mix of antigen receptors changes, therefore, with each individual, increasing the diversity of

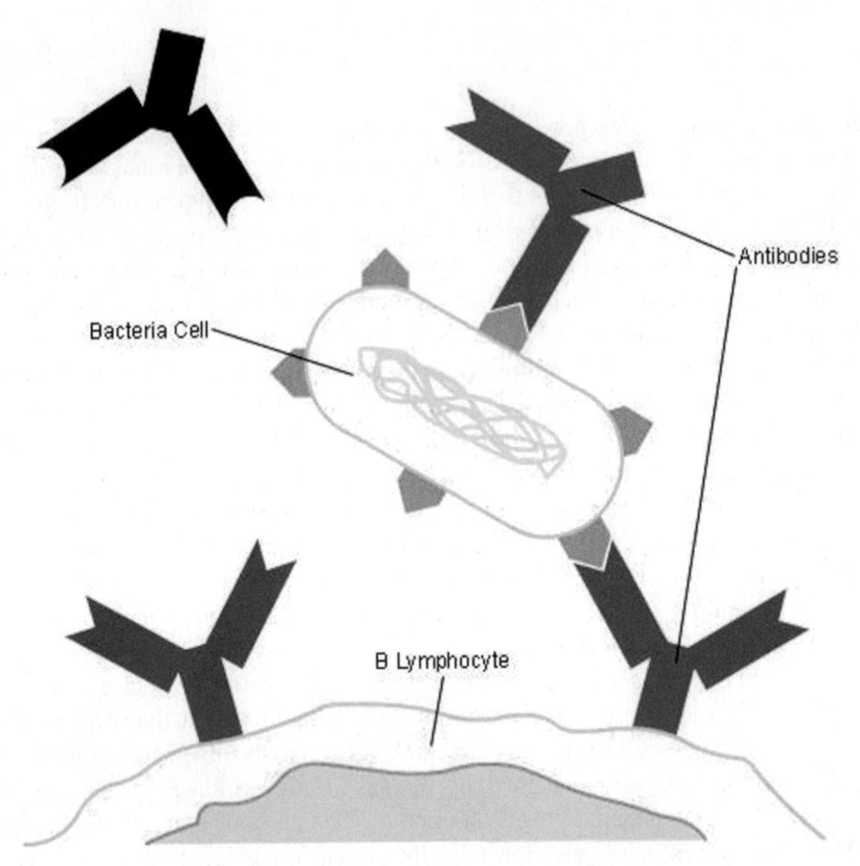

FIGURE 5-3: Activation of a B-lymphocyte

antibodies within a population. As a result of this somatic shuffle, not even identical twins have the same populations of B-lymphocytes.

Usually, an antigen will fit some receptors better than others. In other words, it has a high affinity for some receptors. It may, however, fit others with low affinity. So, a given antigen will actually induce the production of a number of different antibodies, some of which are more effective than others. (See Figure 5-4.) In the body, the population of activated lymphocytes is not *monoclonal* (coming from one clone) but rather is *polyclonal*, coming from a number of different lymphocyte clones. The size of the response of individual clones tends to correspond to the affinity of the antigen for the B-lymphocyte receptor in question. After a lymphocyte is activated and producing antibodies, the B-cell may undergo somatic rearrangement. This produces some new types of cells and allows the body to select cells that have higher affinity for the antigen.

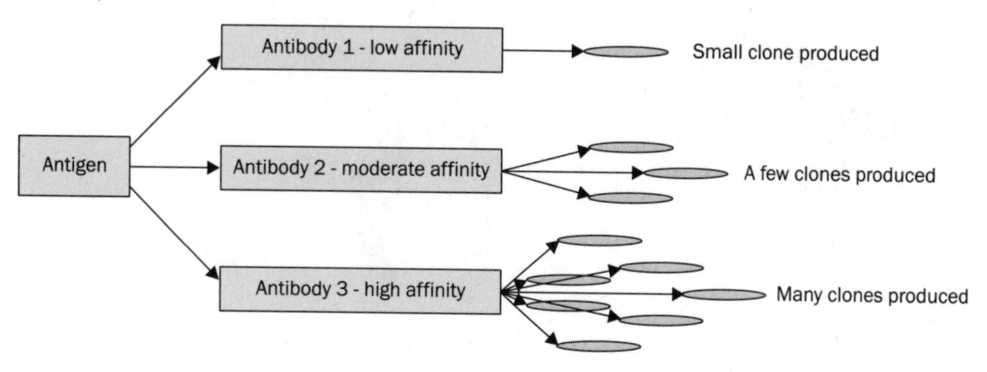

FIGURE 5-4: Lymphocyte clones producing antibodies with different antigen affinity

Structure of Antibodies

Antibodies, as shown in Figure 5-5, are shaped like crabs, with a stalk and two appendages. The body of the crab is the same for all antibodies of a given organism; it's the pinchers that are specialized to fit just one molecular configuration. Each antibody has two identical binding sites, one per pincher.

The appendages are reminiscent of crab pinchers also in the fact that there is one big (heavy) chain and one small (light) chain. At their ends, the chains form loops linked by sugar bonds (*glycosylation*). The receptor site is formed by these loops.

Like much of biological terminology, the names of the parts of the antibody are derived from the early experiments with these compounds. The stalk is known as

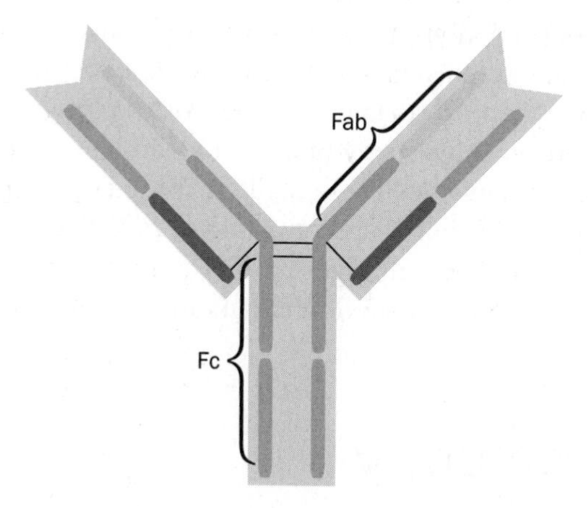

FIGURE 5-5: Basic structure of an antibody

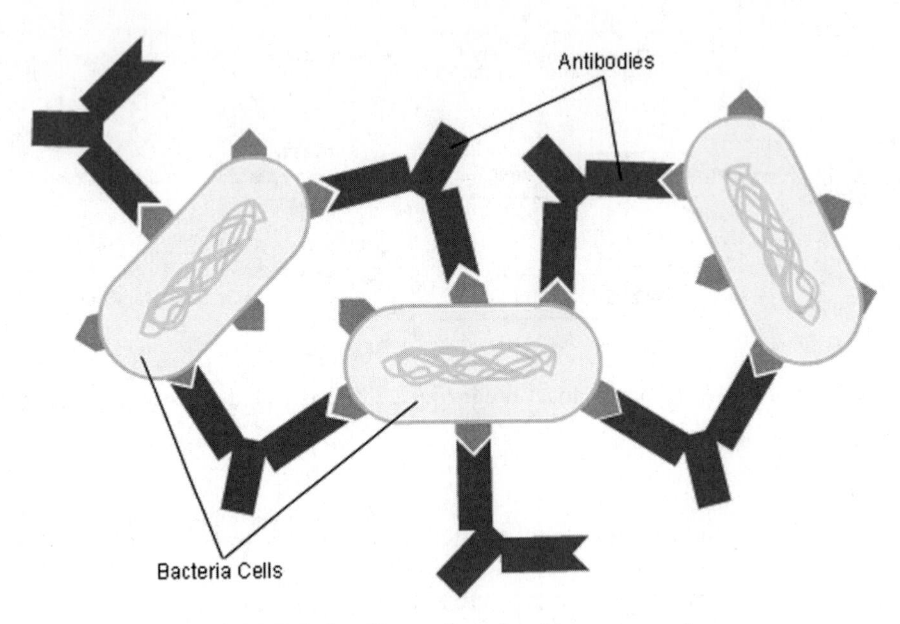

FIGURE 5-6: Formation of an immunocomplex

the *crystallizable fragment* (Fc) because when early researchers cut the antibodies at the hinge regions with enzymes that attack proteins (proteolytic enzymes), they found that one fraction, which turned out to be the stalk, could be made to precipitate out of solution as a crystal. The Fc portion of the antibody increases the life span of the molecule in the blood stream and tissues and initiates important immune reactions at the site of an inflammatory response. However, the presence of the Fc portion of the molecule prevents antibodies from diffusing readily into cells.

The antigen-binding activity is associated with the fragments that do not precipitate out of solution, henceforth known as the *antibody fragment*, or Fab. Because there are two pinchers, each antibody binds to two antigens and each antigen may have more than one binding site. These additional binding sites bind to two additional antibodies, etc. The result is the formation of rather large compounds known as *immunocomplexes*, as shown in Figure 5-6. The formation of the immunocomplex is important in the destruction/elimination of the foreign material.

Cell-mediated Immunity

In some cases, the immune system needs to eliminate material that takes up residence inside cells or are otherwise not available to circulating antibodies. Examples

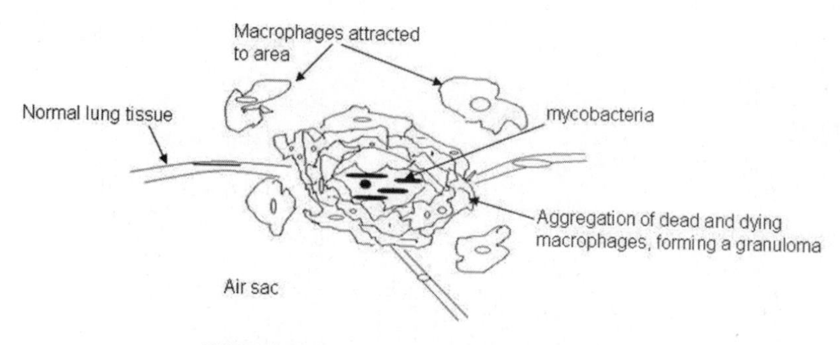

FIGURE 5-7: A granuloma in the lung

are viruses and some kinds of bacteria with waxy surfaces, called *mycobacteria*. An example of the latter is the organism that causes tuberculosis. Unfortunately, the cell housing a foreign invader must itself be destroyed. The destruction is done either by macrophages, natural killer cells (NKCs), or killer T-lymphocytes. The infected cell may be ingested by the macrophages. Alternatively, its membrane may be attacked by an NKC resulting in a loss of the integrity of the barrier.

In the case of the mycobacteria, the bacteria sometimes win. If they are ingested by a macrophage, they may kill the ingesting cell. Other macrophages attempt to wall off the offending organism and may themselves be killed as the bacteria proliferate. The result is a lesion called a *granuloma*. In diseases like tuberculosis, the granulomas can grow in size and number and eventually compromise the ability of the lung to absorb oxygen, as seen in Figure 5-7.

Immune Response to Proteins

The simplest picture of the activation of the immune system is this. The antigen adheres to an antibody, B-lymphocytes proliferate and produce more antibodies, and the antigen is eliminated. The sequence of events works for antigens with highly repetitive structures such as carbohydrates, glyco lipids, phospholipids, and nucleic acids.

The immune response to a protein is more complex and includes steps to help prevent reaction to self-antigens. Mounting an immune response against a protein usually involves a series of handoffs between different types of cells in the immune system. This more complex process is depicted in Figure 5-8.

The development of an immune response against proteins depends on a specific type of receptor on the surface of specific immune cells. This important receptor is called the *major histocompatibility complex* (MHC). The MHC is so named because this particular antigenic receptor appears to be crucial in the immune system's recognition of "self." There are two major types of MHCs. The first type appears on

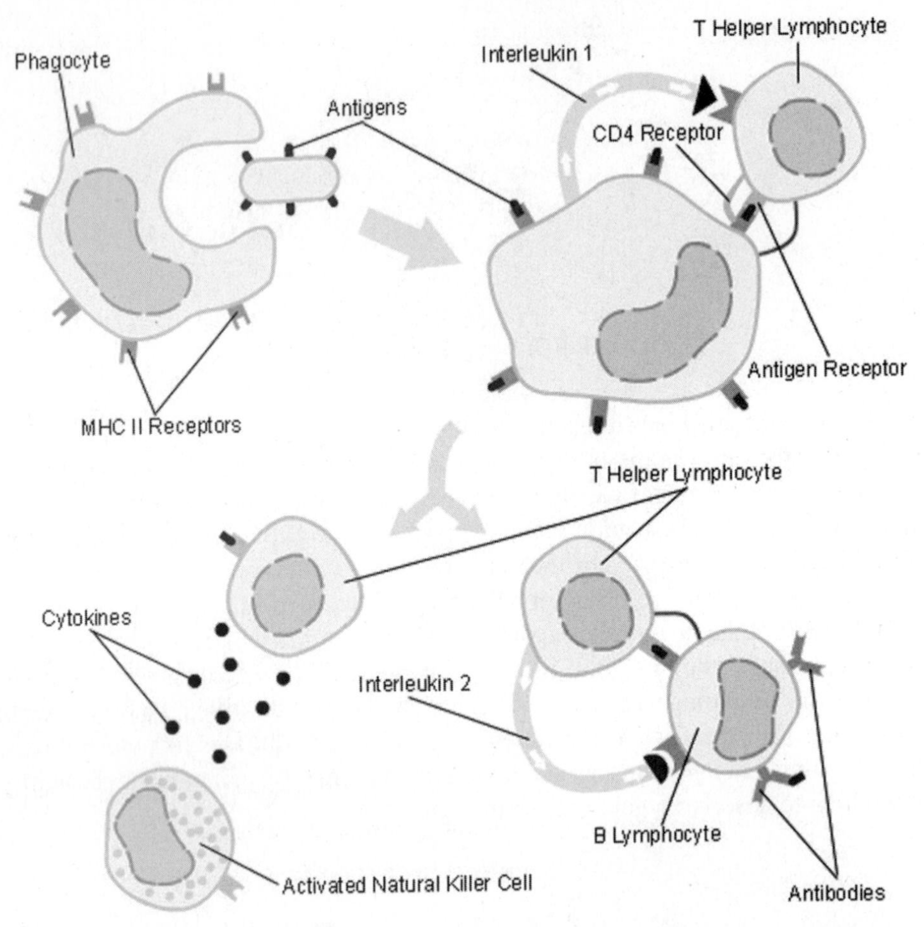

FIGURE 5-8: Schematic process of immune response to proteinaceous antigens

all cells of the body, as the primary "this is me!" flag. These are named MHC Class I. The second type appears only on specialized cells of the immune system and must be present for an immune response to occur; they are cleverly named MHC Class II.

Here's how the immune response process works. When a complex antigen appears in the system, it is ingested by an *antigen-presenting cell* (APC). There are a number of types of cells that can present antigen. The immune response in some cases depends on which type of cell is presenting the antigen. Most APCs are macrophages. In any case, the APC digests the protein antigen into small units (the *peptide*) and then transport the peptides to the surface where it is "presented"—that is, held out for all the world to see. When the T-cell "sees" the antigen and the MHC Class II together, the T-cell is *activated*. The activated T-cell in turn will secrete chemical signals that induce B-lymphocytes to produce antibodies or initiate a

cell-mediated immune response (see the discussion that follows). These chemical signals are called *interleukins*. Note that the name interleukin is pretty general—it just means a chemical secreted by leukocytes, i.e., white blood cells, and affecting leukocytes. The T-cells also activate macrophages, which in turn secrete interleukins that recruit additional T-cells. These additional cells are called *helper* T-cells and they magnify the immune response.

For a T-cell to respond as it should, it must be sufficiently mature. Mature T-cells have a receptor on their surface called CD4. This receptor is responsible for the-recognition of the MHC Class II receptor. Note the quality control checks. The immune system is activated by a cell that requires interface with a cell that has taken in and digested the antigen *and* has an identifier on its surface. The identifier tells the T-cell that not only is the APC "self" but that it is a scout of the immune system and is presenting foreign antigens. This design helps to amplify the immune response and allows additional quality control to prevent reaction against self-antigens.

A special class of APCs is called *dendritic cells*. Dendritic cells are rare and difficult to isolate but they have a very important role. Distinct surface antigens that characterize dendritic cells have only recently been identified. Again, the name has more to do with what early researchers believed about the cell than with what we now know is the function of the cell. Dendritic cells are so called because they are spiny. They have such abundance of long, thin tentacles, that when first observed, they were assumed to be part of the nervous system. Because they mature in the thymus, dendritic cells are by definition T-cells. Dendritic cells are found throughout the body, especially in epithelial tissues of various organs. A special class of these cells is in residence in the lymph nodes, where immature forms of the cells routinely process and hold on their surface (present) self-antigens. As long as the self-antigens are held on the surface of immature dendritic cells, the immune system will not mount a reaction to these antigens. The proper function of dendritic cells is crucial to the prevention of autoimmune diseases.

Dendritic cells are derived from either lymphocytes or monocytes. One scheme of classification identifies the cells derived from monocytes as DC-1 and those from lymphocytes as DC-2. This convention is admittedly a gross oversimplification but it is useful in explaining the role of the dendritic cell at a high level. DC-1 cells present antigens to the T-cell receptor and secrete interleukin 12 (IL-12). The IL-12 (probably numbered that because it was the twelfth discovered) in turn causes the reaction to accelerate by inducting the T-cells to secrete additional chemicals (called *cytokines*). These then excite macrophages to kill internal pathogens, induce NKT (killer T-cells) to attack by creating pores in the target cells, and initiate inflammation. DC-2 cells secrete other interleukins that stimulate B-cells, attract eosinophils, and promote the synthesis of IgE antibodies, a type of antibody associated with allergic responses. AIDS/HIV infects one type of dendritic cell, which then travels to the lymph nodes and moves on to helper T-cells. Note that, unfortunately, the CD4 molecule is a receptor for HIV. The HIV organism binds to the CD4 T-cell receptor and compromises the immune system of the victim.

In sum, response to proteins requires a handoff of the antigens between different cells in the immune system. One of these cells is a reservoir for self-antigens so that the body can compare the self versus foreign antigens. Once the handoff is complete, the cells receiving the antigens communicate using cytokines to cells responsible for the actual response. See Figures 5-8 and 5-9.

Other Components of Immunity

The human body is designed to protect its borders. The nonspecific defenses of the interface between the body and the outside world include saliva in the mouth, enzymes in the digestive tract, acid in the stomach, and the blanket of moving mucous that clears invaders from the lungs.

If a pathogen makes it through the barriers, including the skin, the epithelium of the lungs and the epithelium of the gastrointestinal tract, the body is capable of an immediate, nonspecific response. Bacteria and dying cells release *pyrogens*, resulting in a fever. The higher temperature depletes the blood of iron. Iron is essential for the rapid growth of bacteria. Infected cells release *histamine* and *prostaglandins*. These chemicals initiate a nonspecific inflammatory response by causing local blood vessels to dilate and by increasing the permeability of local capillaries. The inflammatory response, once initiated, creates an environment locally hostile to all life-forms. *Bacteria endotoxins*, a general name for chemicals that are toxic to bacteria, and interleukin-1, which attracts lymphocytes, are released by macrophages after they encounter an invading microbe. Additional cells are attracted to the site, including monocytes that develop into additional macrophages, and other leukocytes that contain highly toxic chemicals, especially neutrophils.

The arrival of antibodies against specific antigens at the site increases the intensity of the fight. The blood contains a complex of about 20 different blood serum proteins, known as the *complement system*. Antibodies contain complement receptors on the stalk or Fc component. The complement proteins are proteolytic enzymes that don't work unless they have a piece chopped off. The first enzyme is activated by binding to the antibody receptor and chops off the inhibiting part of the second enzyme, which chops off the next, etc. The system grows with each step. The activated complement proteins aggregate and form a membrane attack complex. This complex inserts itself into the membrane of the foreign microorganism and forms a pore. The activated complement proteins also attract additional macrophages to the site of infection.

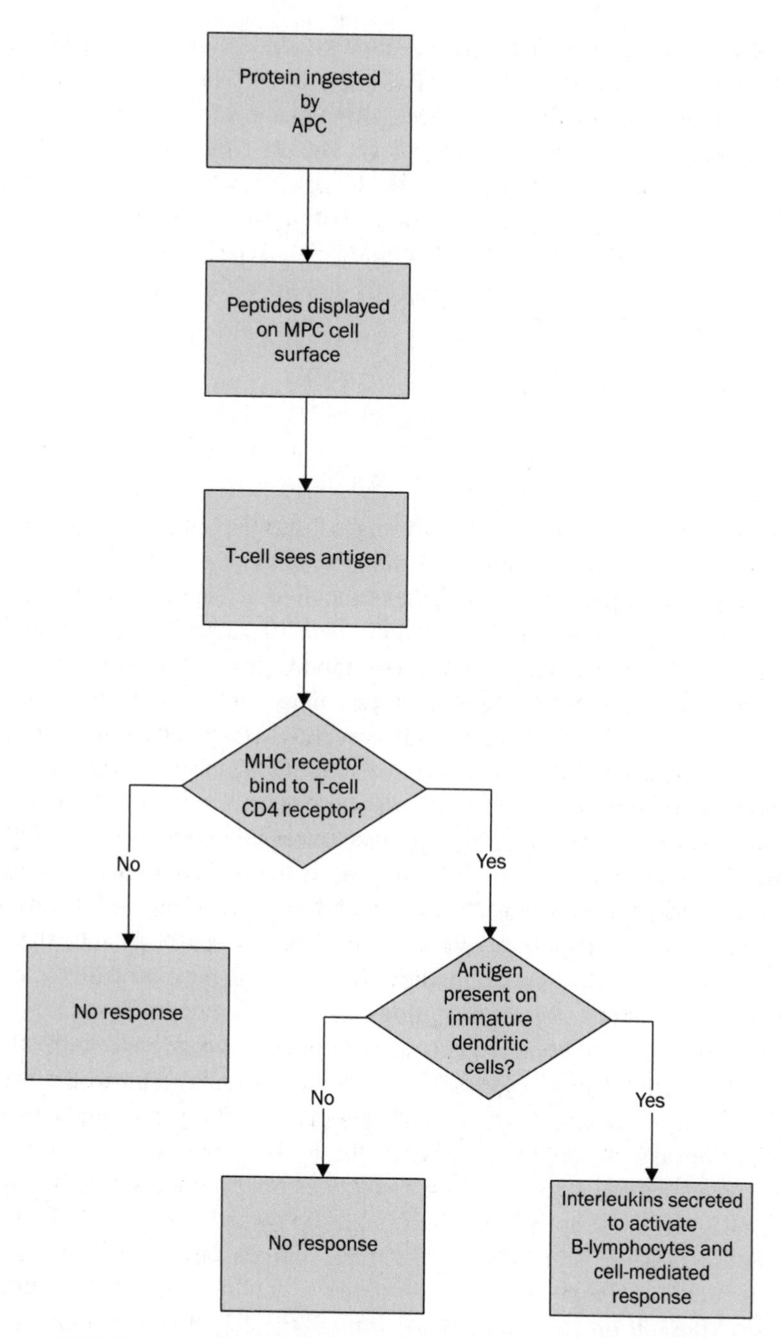

FIGURE 5-9: Flowchart of the process of immune response to proteinaceous antigens

Other components of the nonspecific immune system include cells that patrol the tissues looking for abnormal cells. Natural killer cells (NKCs) are like soldiers, responsible for eliminating cells infected with viruses or potentially dangerous transformed cells. NKC respond to chemicals secreted by infected or abnormal cells and by activation by dendritic cells, as described earlier. These cells bind to the surface of the target and create lethal pores within the membrane.

The nonspecific defenses against viruses include a class of proteins called *interferons*. Interferons are released by virus-infected cells and interfere with the ability of the virus to infect surrounding cells.

Summary

The immune system is like an army with many different branches and units with different specialties. Most cells of the immune system originate in the bone marrow. These include lymphocytes, monocytes (which give rise to macrophages), and granulocytes (which secrete chemicals that kill cells). Lymphocytes that stay in the bone marrow until maturity are known as B-lymphocytes. B-lymphocytes produce antibodies. Each individual has lymphocytes capable of producing hundreds of thousands of antibodies against recognizable structures on various molecules, called antigens. An antigen is simply a structure capable of causing a lymphocyte to produce antibodies. A lymphocyte that has never seen its own specific antigen is called a "naïve" lymphocyte. If this lymphocyte encounters its antigen (and it may not), the antibodies that are sticking out of the surface of the antigen lock onto said antigen and activate the lymphocyte. The activated lymphocyte divides to produce a colony, and the colony begins to crank out the specific antibody into the blood stream and lymphatic fluid. The antibodies bind to and deactivate the foreign protein, primarily by forming large aggregations called immunocomplexes.

Lymphocytes that leave bone marrow as adolescents and migrate to the thymus to mature are called T-lymphocytes. T-lymphocytes are important in the immune response to proteins. The body is very careful about mounting an immune response to proteins because a mistaken response to the body's own proteins would have disastrous consequences. As a result, the body does not produce antibodies against most proteins until (1) the antigen has been taken up by and presented to T-lymphocytes by an antigen-presenting cell, and (2) the antigen-presenting cell has been identified as "self" by the presence of a structure called the major histocompatibility complex (MHC) II on its surface. Furthermore, the T-lymphocyte compares the antigen to antigens on the surface of specialized cells called dendritic cells. If the dendritic cells are immature, then the antigens on their surface are self-antigens—the T-cell will not initiate a response. If the dendritic cells are mature, then the

antigens they present are fair game for the immune system. The T-cells secrete some activating chemicals called interleukins that initiate antibody production by B-lymphocytes, and also activate some marauding cells called natural killer cells. NKCs attack infected cells and also kill mutated, potentially cancerous cells.

The body has nonspecific ways to fight invading microbes. The granulocytes are attracted to infection and secrete nasty chemicals that kill surrounding cells. Macrophages ingest and attempt to wall off invaders. Once specific antibodies arrive, however, the immune response is more effective. The stalk of the antibody molecule binds a big complement protein. Once bound, the complement protein undergoes a modification as pieces of it are chopped off. The final complement protein initiates several chemical reactions that kill the foreign cells.

Quiz

1. Dendritic cells:
   (a) present self-antibodies and remind the immune system not to respond.
   (b) activate T-lymphocytes to induce antibody production.
   (c) are found in the blood stream.
   (d) a and c are true.
   (e) a, b, and c are true.
2. Inflammation:
   (a) is specific to a given antigen.
   (b) is enhanced when antibodies are present.
   (c) requires B-lymphocytes.
   (d) requires T-lymphocytes.
   (e) requires macrophages.
3. The complement system:
   (a) is a type of antibody.
   (b) kills cells nonspecifically.
   (c) is produced by T-lymphocytes.
   (d) consists of proteins normally found in circulating blood.
   (e) b and d are true.
   (f) b, c, and d are true.

4. T-lymphocytes:
   (a) produce antibodies.
   (b) kill other cells.
   (c) activate B-lymphocytes.
   (d) suppress the immune system.
   (e) all of the above.
   (f) b, c, and d are correct.
5. Interferon:
   (a) is produced by B-lymphocytes.
   (b) is produced by macrophages.
   (c) prevents viral invasion of cells.
   (d) a, b, and c are correct.
6. An antigen-presenting cell:
   (a) is another name for B-lymphocytes.
   (b) has MHC Class II receptors on their surface.
   (c) kill virus-infected cells.
   (d) activates the complement system.
7. Monoclonal antibodies:
   (a) are much better than polyclonal antibodies in forming immunocomplexes.
   (b) are antibodies formed against a specific clone of antigens.
   (c) all have identical receptor sites.
   (d) are all produced in the bone marrow.
8. The "stalk" of the Fc portion of an antibody:
   (a) binds antigens.
   (b) binds the first protein of the complement cascade.
   (c) is necessary to extend the life of the antibody.
   (d) allows the antibody to diffuse freely into cells.
   (e) b and c are correct.
   (f) a, b, and c are correct.

9. T-lymphocytes:
   (a) are the target cell of the AIDS virus.
   (b) have a receptor that recognizes the MHC Class II antigen on the surface of antigen-presenting cells.
   (c) are important in mounting an immune response against sugars and lipids.
   (d) mature in the thymus.
   (e) all are correct.
   (f) a, b, and d are correct.
10. Macrophages:
   (a) are derived from monocytes.
   (b) will ingest cells, cell debris, and bacteria.
   (c) will attempt to wall off material they cannot eat.
   (d) all are correct.

CHAPTER 6

Immunotherapy and Other Bioengineering Applications

Immunotherapy is far from new. Physicians have long understood the role of the immune system in general health and in fighting cancer. What is new is the understanding of how the whole system works, the identification of specific antigens on certain cells (including not only tumor cells but on specific cells of the immune system as well) and monoclonal antibodies. These developments have the potential to dramatically improve health care.

Nonspecific Immune Stimulation

Therapists have understood how to nonspecifically stimulate the immune system for several decades. One can provide a patient with a general mixture of *gamma globulins*, or collection of antibodies, which can be isolated from donated blood provided by other individuals. The hope is that some of these antibodies will be helpful in halting an advancing infection. *Interferon*, which nonspecifically impedes viral invasion into cells, is now available from bacterial sources, derived using bio-engineering techniques discussed in Chapter 7. Finally, it is known that the immune response is heightened under certain conditions. For example, injections of antigens in association with oils will increase the response. Materials that will induce this increased response are known as *adjuvants*. Adjuvants appear to present the immune system with danger signals that indicate that things are really bad.

Antisera

Antisera is derived from the blood of animals or individuals who have been exposed to a given antigen. If you are bitten by a snake, you will be given a type of antisera called anti-venom. This serum is derived from an animal, presumably some poor horse, which has been injected with the snake venom at a level less than lethal. The horse generates antibodies against the snake venom. These antibodies are collected, concentrated, and injected into you to bind the venom and prevent its biological effect. Do you, in turn, mount an immune response against the horse antibodies? You betcha.

Antisera contains a mixture of antibodies, some of which bind the antigen more avidly than others. The lymphocytes producing the mixture are *polyclonal*, or derive more than one clone. This is actually preferable to antibodies derived from one clone (*monoclonal*), even if the monoclonal antibodies have a high affinity for the antigen. (See Figure 6-1.) The combination of variant antibodies and antigen form large clumps, known as

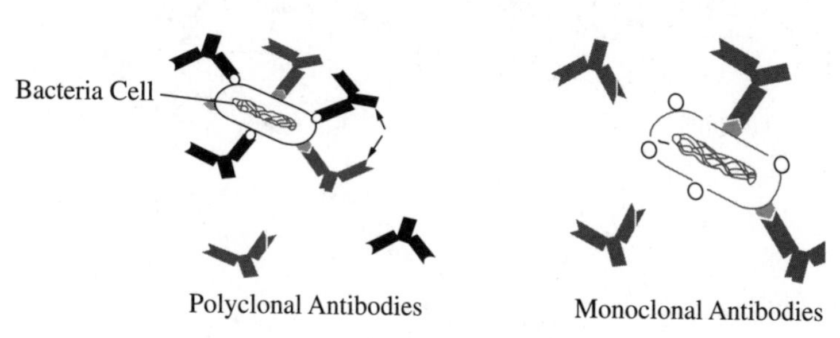

Figure 6-1 Antisera versus antibody

immunocomplexes (described in Chapter 5), that effectively wrap up the antigen. These complexes are relatively insoluble and sometimes precipitate out of the blood stream *in vivo*. *In vivo* means inside of a living organism. The formation of immunocomplexes is used to concentrate antibodies from antisera. After the blood is withdrawn from the animal, such as your horse that is producing antivenom for you, the ratio of antigen and antibody is altered, usually by adding antigen. Immunocomplexes will be formed and will precipitate out of solution. Then the antibodies can be isolated.

Monoclonal Antibodies

Monoclonal antibodies (mAbs) earn their name because they are produced by a single clone of cells. This is a physical phenomenon that you can observe in B-lymphocytes grown in culture. As you know, lymphocytes are stimulated to produce antibodies against a specific target, or antigen, when said target material binds to their surface. The binding not only prompts the lymphocyte to turn out the appropriate antibody but induces cell division as well, resulting in clones of this particular cell. In cell culture, some of the cells challenged with antigen divide and form masses that you can see with the naked eye. These are the clones that are producing a single type of antibody per clone. The proliferation of the single type of lymphocyte that you see in culture also occurs in the body.

There are numerous medical applications for human monoclonal antibodies. Theoretically, these antibodies could be obtained by B-lymphocytes grown in the laboratory. Unfortunately, clones of human B-lymphocytes are very difficult to maintain in culture. It is much easier to expose rats or mice to the antigen of interest and develop clones of lymphocytes producing the correct antibodies from these. Historically, the commercial production of some monoclonal antibodies has used both human and rat or mouse cells in a combination called *hybridomas*, seen in Figure 6-2.

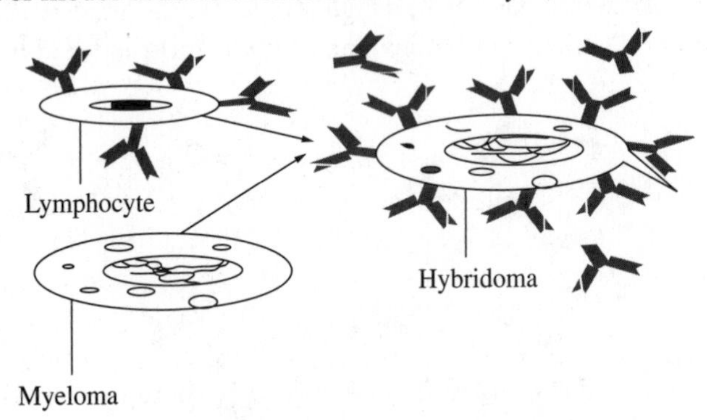

Figure 6-2 Hybridoma

Hybridomas are spleen cells from rats or mice (these animals are called *murine*), that have been fused with human cancerous myeloma cells. The myeloma portion of the cells provides the hybridomas with the ability to divide and grow in culture, theoretically indefinitely. The mouse or rat lymphocyte portion provides the ability to produce antibodies. The fusion can be induced by a virus, an electrical field, or polyethylene glycol. This technique has some disadvantages. The production of antibodies by hybridomas is expensive and time-consuming. Furthermore, the rodent antibodies produced by hybridomas are just fine for analytical work, such as in the ELISA assay described later in the chapter, or in staining tissue samples, but they have limited therapeutic potential in humans because they are rejected as foreign.

Happily, you, as a bioengineer, can engineer antibodies. Theoretically, you can determine the genes that result in the production of any antibody you can find by using the techniques described in Chapter 7. As you will learn, you can decipher the gene producing the antibody you want and place that gene into bacteria. The bacteria will then produce the antibody. In practice, bacteria-produced antibodies have limitations. It turns out that the bridges between the chains on the antibody are important, and bacteria cannot perform the required chemistry to produce the bridges.

For many monoclonal antibodies, bioengineers have turned to another type cell that produces antibodies in the form needed. Chinese hamster ovary (CHO) cells have been a great hit. The CHO cells are the host of choice for current production of bioengineered monoclonal antibodies. The major drawback for the CHO cells is that these antibodies still have antigens that the human body recognizes as foreign, causing them to be rejected.

The advent of genetic engineering has provided another approach that has greatly increased the therapeutic potential for monoclonal antibodies. Using recombinant DNA techniques, rodent cells—such as the CHO cells—can be engineered to produce antibodies with a human stalk. Most of the antigens that will cause a non-self antibody to be rejected are on the stalk. Antibodies produced this way are called *chimeric antibodies*. Chimeric antibodies dominate current usage in immunological and oncological (cancer) treatments. Recently, mAbs have been further "humanized" by genetically replacing the majority of the rodent structures on the Fab with human ones. Even more recently, fully human mAbs have been developed, brought into trial, and a limited number put in therapeutic use. Of the four types—rodent (murine), chimeric, humanized, and human—the chimeric is the largest number in use, but human is the largest number in current clinical trials.

There are 18 therapeutic mAbs currently approved by the FDA, most of which are either rodent (murine) or chimeric. The naming conventions for these mAbs are as follows. All monoclonal antibody names end with "mab." If the antibody is from a mouse, the letter o is added to become "omab." The chimeric type adds the letters xi to become "ximab," and the humanized type adds "zumab." For example, the popular immunotherapy drug Herceptin, used to treat breast cancer, is actually trastuzumab. Therefore, it is a fully human monoclonal antibody. Rituxan, used to treat lymphomas, is actually rituximab, a chimeric monoclonal antibody.

The major current uses are (1) immunosuppression for autoimmune disease victims and for organ transplant recipients, and (2) cancer treatment. Monoclonal antibodies for immunosuppression target receptors on the surface of lymphocytes. Monoclonal antibodies for cancer treatment target receptors found predominantly on cancer cells.

The science is too new to achieve large inroads into therapeutic practices yet, but be aware, changes are coming. From modest beginnings in the early '80s, the current monoclonal antibody (mAb) market is growing at a rate of about 20 percent a year. Most development work is being done by small biotechnology companies, but as the profits grow, major pharmaceutical companies are showing increased interest.

Some of the major side effects with the use of monoclonal antibodies are related to the initiation of a general inflammatory response. As you remember, a receptor on the Fc fraction of the antibody binds complements and touches off nonspecific inflammation. One approach to this problem has been to administer only the Fab fraction of the antibody, but this form doesn't survive very well in the body. However, you, bioengineer, can do really clever things with your ability to engineer antibodies. You may be able to eliminate the receptors that initiate the complement cascade, and thereby enjoy the advantage of having the stalk, the main advantage being that the life span of the antibody *in vivo* is greatly increased, without the inflammatory response.

Nanobodies

I don't want you to think that all of biotechnology is serendipitous, but...an observant research group working with specific species of camels and llamas has discovered that these creatures possess a very strange form of antibody called *nanobodies*. These entities appear to be small parts of antibodies. Some forms are heavy chains without light chains, but with the antigen affinity of light chains. Alternatively, other forms are light chains with the ability to commandeer other parts of the immune system, a function usually associated with the heavy chain. If a system can be devised to "humanize" the antibodies—that is, to engineer a large enough portion of the protein to fool the human immune system into accepting it—use of nanobodies may overcome some of the obstacles that compromise monoclonal antibody therapy. Nanobodies are much smaller than other types of monoclonal antibodies and penetrate into tissues much more easily. They would be much more effective, for example, in delivering toxins that you can chemically link the antibodies to cancer tissues.

A Library of Antibodies

A large variety of monoclonal antibodies has been and is being developed, as the specific genes that code for specific antibodies are deciphered. A strategy to maintain this large collection of genetic information in one place would be invaluable. To this end, the genes for the light chain of a number of monoclonal antibodies have been introduced into living *bacterial phages*. Bacterial phages are viruses that infect bacteria. The phages are then maintained as a kind of library of this information, a living library that contains the genetic information that you want to keep, just like you would keep a book collection. Such living libraries are known as *combination libraries*.

Theoretically, you can go to the antibody library and order an antibody. The specific phage that has housed the genetic information for your antibody will be provided to you. You can introduce this genetic sequence into bacteria by allowing the phage to invade your bacteria. If you do this correctly, the infected bacteria will produce the antibody light chain that you want. This must be later hooked to a heavy chain to produce the Fab segment of interest. See Figure 6-3.

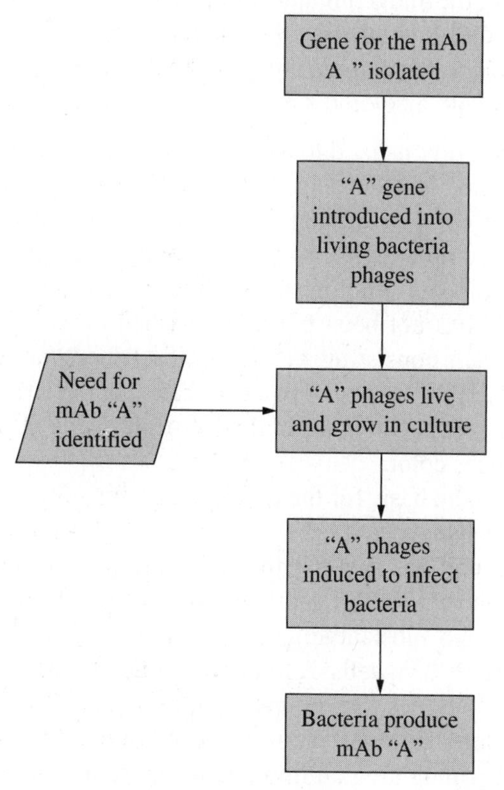

Figure 6-3 How bacteria phages serve as living libraries

ELISA

The development of monoclonal antibodies vastly improved our ability to diagnosis a number of human diseases. Using antibodies against a very specific target, we can detect low levels of antigens or of antibodies that are present in the human body under disease conditions. The presence of the antibodies is a very sensitive indicator that the antigen is present. The antigen might be associated with bacteria, a virus, or an element of the person's own body, against which the immune system is inappropriately mounting a defense (autoimmune disease).

Consider this challenge. You as a diagnostician want to know if your patient has a certain antigen in his/her system. You might suspect a streptococcus infection and want a quick screen. You know that a patient with a strep infection would have strep antigens. A very widely used monoclonal antibody technique is called ELISA (Enzyme-Linked Immunosorbent Assay). ELISA is a method of detecting the formation of an immunocomplex, the linkage of antibodies and of antigens. You will use the following:

- An antibody to the antigen in question, attached to a substrate
- The patient's blood, containing the antigen
- A free antibody to the antigen in question; this one with a linked enzyme that will react with a colored substrate

In ELISA, the antibody is fixed to a plastic well and picks up whatever antigen is present in the patient's blood, as shown in Figure 6-4. Then the material is treated with another copy of the antibody, this one enzyme linked. The bound antigen will form a complex with the new antibody by binding the new antibody at a site other than the site binding the antibody fixed to the well. The enzyme-linked antibody that has not been bound is washed away. The remaining molecules form a sandwich consisting of layers of antibody/antigen/enzyme-linked antibody. This sandwich is treated with a substrate that the enzyme will digest. The digestion results in a change in color in the substrate. Voilà! The antigen is present if the substrate changes color.

This technology is the basis for the quick screen for strep infections and for the over-the-counter pregnancy-test kits.

There are other conditions in which the antigen may not be circulating or may be present in the body in very small quantities. An example is HIV infection (or other viruses) where, during a substantial period of the infection, the antigen is sequestered inside the victim's own cells. Another example of low-level antigen presence is the need to detect antigens for victims of autoimmune diseases. You might have a child patient at risk for diabetes and need to determine if the child has antibodies that indicate he/she is mounting an immune response to his/her insulin-producing

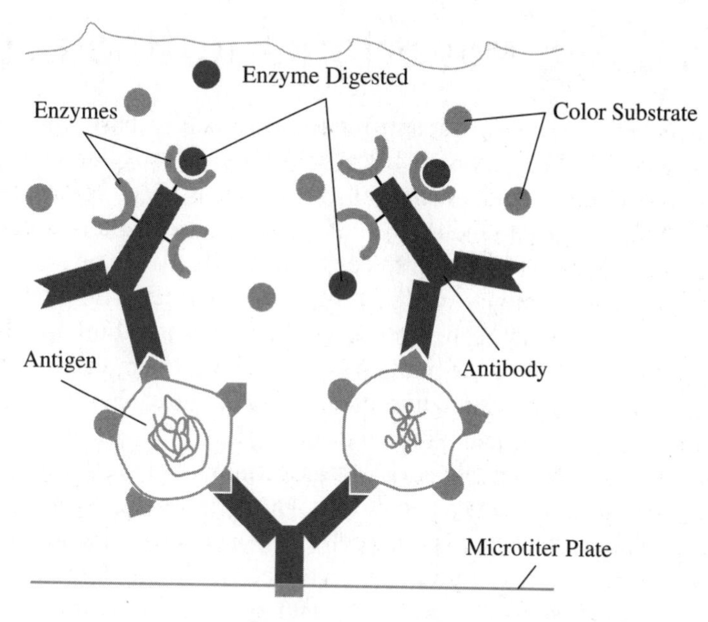

Figure 6-4 ELISA

cells. Remember, once activated by exposure to an antigen, B-lymphocytes continue to produce antibodies, although at decreasing levels. The detection of circulating antibodies is a very sensitive test for the presence of the antigen somewhere in the body.

In this case, you would use a modification of ELISA, whereby the antigen is attached to the substrate, and the patient's blood containing the antibody is allowed to react with the antigen. To detect the formation of this immunocomplex, you will need another antibody with an enzyme link. This antibody is engineered to react to any human antibody. It is an anti-human antibody. This antibody attaches to the antibodies that are attached to the antigen simply because they are human antibodies. The excess is washed away. Where the second antibody has attached, the enzyme digests the substrate so that you can see the immunocomplex. It's another sandwich!

One note of caution: most assays need to determine the trade-off between sensitivity and accuracy based on the use of the results. In the case of HIV, you may want to make sure that you don't tell patients that they are HIV negative when in fact they are HIV positive. As a result, you would rather err on the conservative side even though sometimes the assay will determine that a patient is HIV positive when in fact he or she is not. In other words, the assay for HIV is very sensitive but is prone to false positives. Final determination of the presence of HIV therefore requires a repeat of the test or the performance of additional, more accurate tests.

Other Assays Using Monoclonal Antibodies

Immunocytochemistry is used to identify parts of a cell in basic research and in diagnosing a disease. Basically, if you can make an antibody for something, you can find it in blood, in a tissue, or in a cell. The tissue is sliced or diced as desired and bathed in the antibody to the antigen of interest. This might be a cell receptor or another specific protein product of the cell. The antibody can be linked to a fluorescent dye so that the antigen can be visualized. When the antibody finds and forms a complex with the antigen, the free antibody is washed off and the fluorescent dye is found only where the antigen is located. Using an instrument called a *spectrophotometer* that locates the fluorescent dye, the researcher can determine if and where the antigen is located within the cell.

One commercially viable application of this technique is in cell sorters. Consider this theoretical scenario. You are a pathologist who is trained to recognize the subtle changes in morphology that characterize cells from a victim of rheumatoid arthritis. To find these cells, historically you would collect blood samples, concentrate the cells, prepare stained slides of these cells, and spend hours scanning the slides, looking for the telltale cells. Think of the time and money that would be saved if you could mark these cells with a sign that a computer could recognize and then let the computer scan more cells than you, with your human limitations, could possibly do. We have just described a cell sorter.

Cell sorters that can sort cells based on cell size and granularity have been available for a while. These sorters use laser detectors and computer technology to analyze the laser signals. To detect your rheumatoid arthritis cells, you will need to find a unique antigen associated with these cells and develop a monoclonal antibody against this antigen. Once you have the monoclonal antibody, you can link a fluorescent dye to the antibody. Mix the suspect blood with the labeled monoclonal antibody and send the cells through a cell sorter that has the detector for the fluorescent dye, as in Figure 6-5. You are there!

A *ligand* is a binding agent that can be used to isolate specific molecules from solution. Specific antibodies are perfect ligands. You can use antibodies on a solid matrix support to bind (purify or remove) an antigen. For example, consider a researcher who wishes to study a protein associated with certain types of lymphoma. You can isolate this protein from the blood of lymphoma victims by preparing a long column of beads coated with the antibody to this protein. If you pour the blood through the column, the antibody will bind the antigen. You can use chemical techniques to cause the antigen to be released and thereby obtain a pure sample of the molecule of interest. Antibody techniques can be used to detoxify solutions or remove contaminants. You can also use anti-immunoglobulin antibodies to purify antibody solutions.

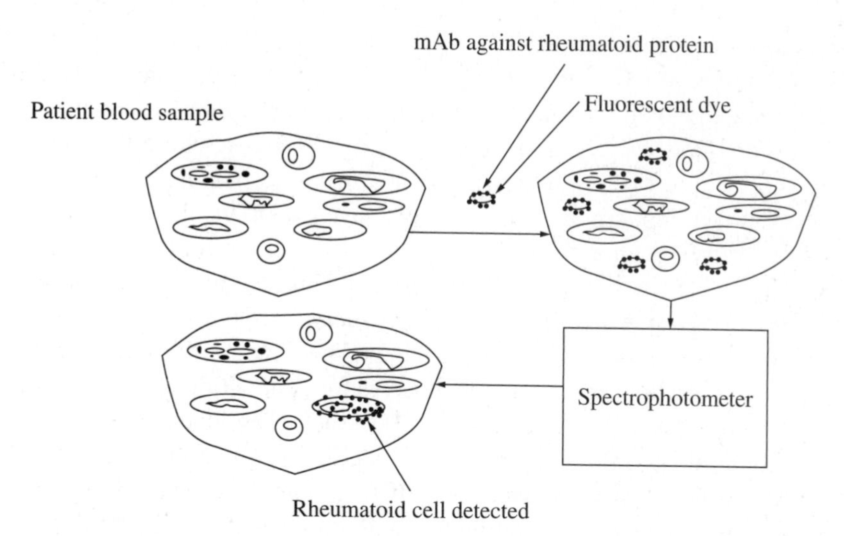

Figure 6-5 Use of fluorescent-labeled monoclonal antibodies to detect rheumatoid arthritis cells

Some experimental *in vivo*–imaging techniques use monoclonal antibodies, such as for imaging of certain types of tumors. Labeling of the antibody with a high-energy but short-lived radioisotope can be used. The labeling involves incubating the cells producing the antibody with a compound that is needed in the antibody structure and is available in a radioactive form, or isotope. The radioactive antibody is injected into the patient. An instrument called a *gamma camera* is used to photograph the radioactive tissue in the living patients. For these applications, the ability of the antibody to rapidly defuse into the tissue is important, so the use of Fab fragments rather than the entire antibody is tempting. Also, the Fab fragments don't have the complement receptor, so they are less likely to initiate an inflammatory response.

Autoimmune Disease

At one time, it was believed that all B-lymphocytes capable of mounting an offense against the body were destroyed early in the development of the immune system. However, obviously, the memory of "self" sometimes fails, and the soldiers turn upon the mother country. One theory as to why this might happen revolves around the fact that many antibodies have cross-reactivity—that is, they have great affinity for a specific antigen but somewhat less affinity for a number of others. Perhaps a virus or another pathogen stimulates production of antigens that have a moderate or very small affinity for self-antigens. We have mentioned that once the immune response is initiated, a process seems to exist that results in the selection of B-lymphocyte colonies with stronger affinity for the antigen of interest. Once this process is initiated, you might actually evolve stronger and stronger reactions against yourself.

What to do about this? Classically, autoimmune diseases have been treated by generic suppression of the immune system. Corticosteroids have been especially popular. However, modern insights and tools have lead to the development of more specific treatments. Although these treatments are still in clinical trials, they hold hope for the recalcitrant diseases. One approach has been to develop antibodies to antibodies. For example, a disease called *myasthenia gravis* is due to the production of antibodies against the acetyl choline receptors on the muscles. See Figure 6-6. Acetylcholine must be taken up by the muscles for contraction to occur. Without it, the muscles lie still and eventually atrophy. If the acetyl choline receptors are bound up by the antibodies to these receptors, the needed acetyl choline cannot be taken into the cell. Antibodies that attack the self-antibodies have been developed. It is hoped that these bioengineered antibodies will bind the self-antibodies before they can attack the acetyl choline receptors on the muscle cells. The antibodies against "self" will be bound up harmlessly in immunocomplexes.

Other treatments are focusing on the dendritic cells. As we have seen, continuous display of self-antigens by immature dendritic cells appears to be essential in the prevention of autoimmune disease. This observation has lead to some clinical trials where dendritic cells are stimulated to take up self-antigens and reintroduced to the host. The challenge here is to prevent the dendritic cells from maturing; once mature, they will actually enhance the immune response. One possibility is genetic engineering for overproduction of proteins that serve to moderate the immune response and inhibit dendritic cell maturation (by inhibition of MHC Class II expression).

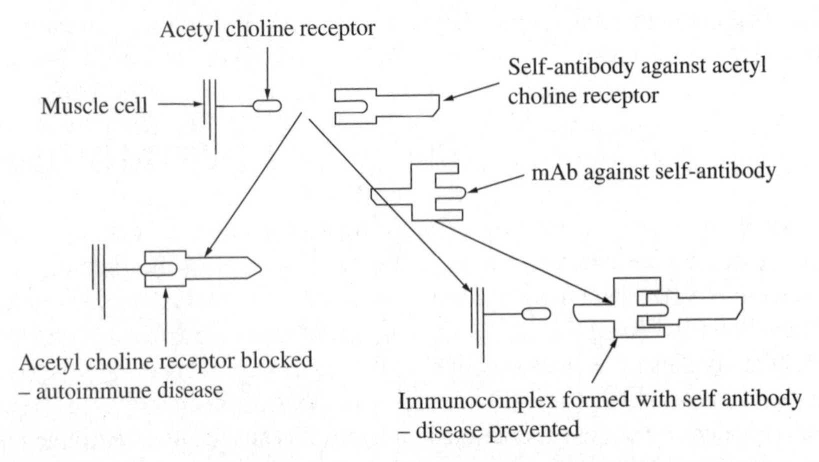

Figure 6-6 Use of monoclonal antibodies in the treatment of myasthenia gravis

Also, bioengineering techniques are being developed that can hopefully prevent gene expression. B-lymphocytes that have the genes to produce self-antigens may be selectively made ineffective by these techniques. (See Chapter 10.)

Therapeutic use of mAbs has not been a risk-free panacea. You can see why effective immunosuppression might increase susceptibility to disease. In fact, use of immunosuppressant antibodies has allowed the development of unusual infectious diseases. One well-known case involves the use of the monoclonal antibody called natalizumab in treatment of multiple sclerosis. A small percentage of patients developed a rare viral disease known as progressive multifocal leukoencephalopathy (PML). The theory is that the mAb, which prevents the normal homing of lymphocytes to sites of infection, blocks the migration of immune cells to the brain, allowing expression of the normally dormant virus. Other problems with autoimmune therapy have been seen with antibodies that block tumor necrosis factors or other *cytokines*—chemicals secreted by one cell that influence the behavior of other cells. These blockages have a negative impact on the immune system in general.

Cancer

Stimulation of the immune systems has long been important in the treatment of cancer. To survive and grow, cancerous cells must overcome the natural surveillance of the immune system. In the 1950s, Dr. William Coley showed he could control the growth of some cancers with an injection of a mixed vaccine to strep and staph bacteria. His concoction is known as Coley's toxin. A nonpathogenic mycobacterium, known as Bacillus Calmette-guerin (BCG), stimulates cell-mediated immunity. It is used as a vaccine against tuberculosis, as well as a treatment for bladder cancer.

Cytokines important to immune function are also used nonspecifically. Recall that interferon is secreted by virus-infected cells to impede the invasion of adjacent cells. You can now produce interferon in large amounts by using bioengineering techniques. One form of interferon (interferon-alpha) is commonly used for some myeloma, myelogenous leukemia, hairy cell leukemia, and malignant melanoma. Recall that interleukins are chemicals secreted by one leukocyte (white blood cell) to stimulate other leukocytes. One type, interleukin-2, is used to treat kidney cancer and melanoma. Unfortunately, these therapies, especially the interleukins, have side effects, usually flu-like symptoms.

Inducing the victim's immune system to attack tumors is difficult because tumor antigens are tolerated by the immune system because these antigens are self-antigens. One strategy is to induce the patient's own dendritic cells to re-present the tumor antigens as foreign. Circulating dendritic cells are removed from the patient's blood and mixed with kidney cancer cells in a laboratory. The cells are then injected back into the patient as a vaccine. These cells have been therefore primed to present the antigen to antibody-producing cells and augment an immune response. Dendritic cell vaccines have been shown to be effective in treatment of some kidney cancers. This treatment regiment has also shown promise in melanoma trials.

The advances in monoclonal antibody therapy have fueled the search for the mythical "magic bullet"—these are cytotoxic weapons that will seek and kill only the cancer cells. The magic bullet continues to evade us. This may be because we are using only one weapon in an arsenal of weapons, or it may be because tumors are diverse, hardy, and hard to kill. Also, tumors tend to be unique. With many tumors, each cancer victim would need his or her own unique antibody. However, laboratory studies have shown that vaccination with lymphoma-associated proteins can stimulate the immune systems of mice to resist the development of lymphomas. There have been some clinical successes in people using the protein vaccines, especially with leukemias and lymphomas. Other immunotherapies include the use of a monoclonal antibody against cell-surface antigens. Examples include rituximab (Rituxan), used in the treatment of non-Hodkin's lymphoma and directed against the CD20 antigen found on the surface of both normal and malignant B-lymphocites. A monoclonal antibody against a receptor found predominately on breast tumors, trastuzumab (Herceptin), is useful against certain breasts cancers.

Again, the use of mABs has experienced some complications due to the relatively nonspecific nature of their actions. Herceptin is active against the HER-2 receptor that is overexpressed on some but not all breast cancers (see Figure 6-7). HER-2 is a growth receptor and its presence may be linked to the abnormal growth patterns of the cancer tissue. However, HER-2 receptors are also present on some normal tissues. A relatively high frequency of cardiovascular complications has been reported in the patients receiving Herceptin. The mABs may be attacking the HER-2 receptors on normal heart cells.

Clinical trials are ongoing to study the effectiveness of linking toxins and antibodies that are directed at cancer cell antigens. The toxin is released when the antibody attached to the cancer cell antigen, killing cells in the vicinity. While this

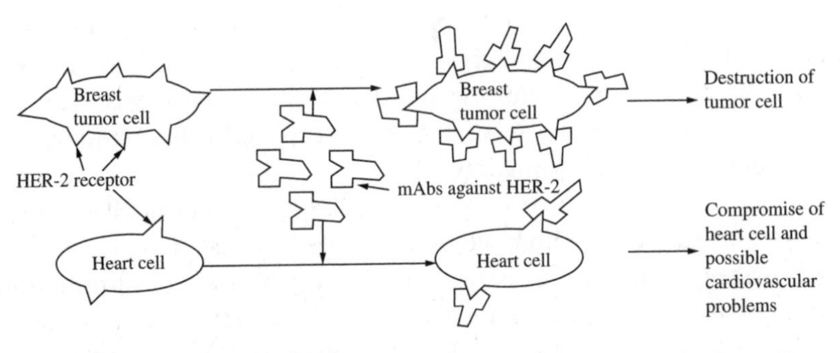

Figure 6-7 Action of Herceptin against breast tumors and other cells

certainly helps to localize the toxin to the area of the tumor, other tissues are affected as well when the toxin is released. The toxin can be linked to the Fab fragment of the antibody. The Fab fragment diffuses much more efficiently into the tissues than does the entire molecule. However, while the Fab fragment can diffuse into the tissues more readily than the intact antibody, the Fab fragment without the stalk has a very short half-life.

The use of immunotherapy in cancer treatment is certainly going to grow. One must appreciate that fact that development of cancer treatment regimens is really expensive and it takes a long time to get even one treatment through the system. However, the number of clinical trials utilizing immunotherapy foreshadows huge advances in this area in the future.

Addictive Disorders

For the imaginative, antibody therapy has limitless applications. One intriguing potential is in treatment of addictive disorders. An antibody can be produced against chemically active, addictive substances. As long as the user of the substance is receiving the antibody against the substance, he or she cannot experience the effect of the drug. For example, an antibody against nicotine is under development and promises to be important in helping smokers fight their addiction to nicotine.

Allograft Rejection

Organ transplant is increasingly common to replace hopelessly diseased organs. Literally millions of patients await availability of an organ that is "compatible," in other words, one that is close enough to the patient in the organ's surface antigenic structure that the immune system does not mount an overwhelming response and shut down the foreign organ. Such organs are called *allografts*. However, even with closely compatible tissues, organ recipients need to be given therapies to nonspecifically suppress their immune response.

Research is underway to improve allograph survival by manipulating the patient's dendritic cells. As discussed earlier under autoimmune diseases, the body can be fooled into accepting a protein as "self" if the protein is presented by immature dendritic cells. The allograft antigens are incubated with the patient's immature dendritic cells. When these cells are reintroduced into the patient, the immune system is fooled into thinking the allograft antigens are "self." Other therapies prevent presentation of allograft antigens to naive T-cells by blocking the receptors on the T-cells that are important in the cells' activation.

Summary

The governing principles of the immune system are fairly simple. There is a large preexisting set of templates that match the majority, if not all of the foreign antigens the body is likely to encounter. If the antigens match a template or receptor on a B-lymphocyte, production of antibodies to that antigen is kicked off. The specific B-lymphocyte population proliferates and stays relatively high to defend against the antigen in question. If the antigen is protein, a system is in place to allow the body to recognize this protein as foreign and yet leave untouched the body's own proteins. This system involves a handoff between antigen-presenting cells, T-lymphocytes, and finally the B-lymphocytes. Autoimmune diseases result when this system breaks down and the immune system attacks sites in the body itself.

In addition to the antibody protection, the immune system also has a patrolling group of cells that recognize abnormal tissue or tissue infected with pathogens. These cells can destroy the tissue by ingesting it or creating holes in its membranes. A nonspecific cell-killing system is touched off in areas of inflammation when the complement proteins are activated and neutrophils release their toxic contents.

The ability to produce specific antibodies commercially has huge medical potentials. These include antibodies that attack cancer cells, antibodies that tie up antibodies in autoimmune diseases, antibodies that damp down the immune system (useful in autoimmune disease and for organ transplant patients), and antibodies that can augment the immune system. While none of these therapies are free of side effects, the large number of clinical trials in this area portent of a major change in medicine.

Quiz

1. Challenges in developing antibodies to tumors:
   (a) tumor cells have no antigens.
   (b) tumor cells may share antigens with other tissues.
   (c) tumor cells tend to be tolerated as "self."
   (d) b and c are true.
   (e) all are true.

2. ELISA:
   (a) is a method for cell sorting.
   (b) is a quantitative method for finding specific antibodies in a sample.
   (c) is a method for cloning lymphocytes.
   (d) is a method for analyzing the antigenic potential of a molecule.
   (e) b and d are correct.
3. Chimeric antibodies are:
   (a) from bioengineered animals.
   (b) from human lymphocytes.
   (c) are part human, part mouse.
   (d) a and c are correct.
4. Allograft rejection:
   (a) is due to production of antigens against "self."
   (b) is due to production of antibodies against foreign antigens.
   (c) is reduced by suppressing the immune system.
   (d) is reduced by enhancing the immune system.
   (e) a and c are correct.
   (f) b and c are correct.
5. Nonspecific immunotherapy would include:
   (a) use of bioengineered antibodies.
   (b) injection of interferon.
   (c) injection of an adjuvant.
   (d) injection of gamma-globulins.
   (e) a, b, and c are correct.
   (f) b, c, and d are correct.
6. Successful cancer therapies have included:
   (a) finding antigens that are only exhibited by cancer cells and developing antibodies against these.
   (b) removing dendritic cells from patients, exposing these cells to tumor antigens, and then reintroducing them into patients.
   (c) removing lymphocytes from patients, exposing these cells to tumor antigens, and then reintroducing them into patients.

(d) poisoning tumor cells by linking poison onto tumor antibodies.
(e) a, b, and d are correct.
(f) a, c, and d are correct.

7. Antibodies can be obtained by:
   (a) immunizing animals and collecting antisera.
   (b) cultivating lymphocytes and exposing the culture to the antigen.
   (c) deciphering the genetic code for the antibody and producing it through bioengineering.
   (d) all are correct.
   (e) a and c are correct.
8. Spleen cells were fused with myeloma cells because:
   (a) myeloma cells are immortal and will grow indefinitely in culture.
   (b) spleen cells don't produce antibodies but myeloma cells do.
   (c) spleen cells cause cancer.
   (d) all are correct.
9. Cell sorters can:
   (a) differentiate between cells based on laser analysis of cytochemical dyes.
   (b) differentiate between cells based on size.
   (c) differentiate between cells based on granularity.
   (d) differentiate between cells based on surface antigens.
   (e) all are correct.
   (f) a, b, and c are correct.
10. Autoimmune disease:
    (a) is due to an individual's immune system turning on itself.
    (b) is treated by general immunosuppression.
    (c) involves the dendritic cells.
    (d) all are correct.

CHAPTER 7

Recombinant Techniques and Deciphering DNA

You now understand at a simplified level how the cell information system works. You, as a bioengineer, are interested in changing it, I am sure. It has undoubtedly occurred to you that, because all living systems use the same information code for proteins, you could induce any cell to produce any product, provided you know the code. This is pretty close to true.

We have discussed why prokaryotic cells are nice to work with. They are relatively easy to maintain in the lab or factory—after all, they are used to living on their own. Their genome is relatively simple. And the majority of their genome is devoted to coding for proteins. You are about to learn how to make a bacteria a protein factory for a protein you desire. Two success stories of this are human insulin and human growth factor. Both of these products came from animals before bioengineering methods were available. They were expensive to produce and were *not* human, so they carried the risk of inducing an immune response and were dependent on an animal source. In this chapter, you will learn how to make your very own human proteins using bacteria.

Recombinant Techniques

To provide a bacterial source for the human proteins you need, you must

- Produce a copy of the DNA sequence that codes for the protein of interest.
- Somehow get this DNA sequence into the bacterial cell.
- Induce the bacteria to produce the protein coded by your DNA sequence.
- Protect your new molecule from the hostile environment of the inside of the bacterial cell.
- Purify your molecule.

These processes will be explained in detail in this chapter. We will use the example of insulin to illustrate how a human protein could be made by bacteria.

PRODUCE A COPY OF THE DNA SEQUENCE FOR YOUR PROTEIN

First, you need to produce a copy of the DNA sequence that codes for your protein. The easiest way to do this is to procure the messenger RNA (mRNA) coding for the gene. As you recall, mRNA is the molecule that transcribes the DNA code for a given protein. The mRNA serves to pick up the code and carry it outside of the nucleus to the ribosomes, where proteins are produced. You can highjack the messenger and get your code. Furthermore, in the eukaryotic cell, by the time the mRNA leaves the nucleus, it has been appropriately snipped and trimmed and rid of any superfluous genetic material that may have been transcribed with the meaningful code, a process called *gene splicing*. If you start with the messenger RNA instead of the DNA itself, you avoid the complexities of DNA transcription, including the variables of gene splicing. Also, a given piece of DNA code is a blueprint that can be used over and over again to produce many copies of mRNA for a given protein. In a cell that is busily making a certain protein, there will be a fairly large number of mRNA's for that protein in the cell cytoplasm. So, use of the mRNA "amplifies" the gene. This is step 1 of the process presented in Figure 7-8.

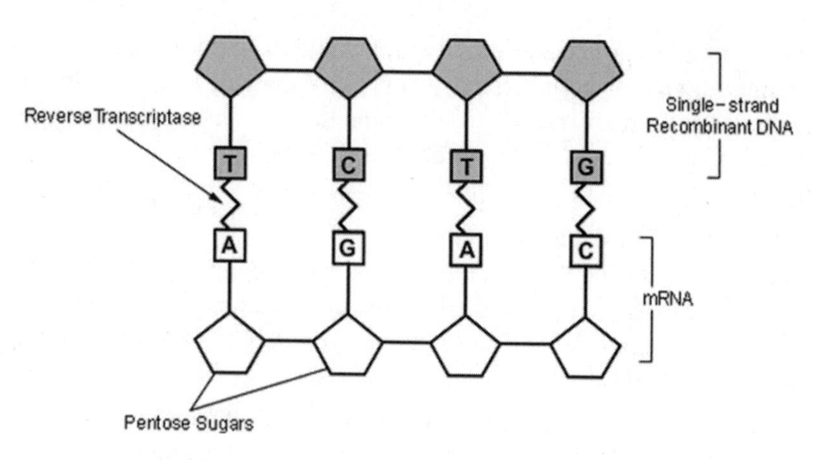

Figure 7-1 Making complementary DNA

To obtain the correct mRNA, you will go to the cell that serves as the normal source of the protein for the body. Consider that you want to make insulin. The source of human insulin is the islet cells of the pancreas. These cells are a production factory for insulin. Molecules of mRNA coding for insulin should be produced at a high rate in these cells. Most of the mRNA in the cell cytoplasm should be for insulin. If you isolate the mRNA from islet cells, you are sure to obtain a large percentage of mRNA molecules that code for insulin. You may wonder how you obtain the mRNA from a concoction made of pureed islet cells. You can isolate mRNA because of RNA's have a chemical signature, a unique poly (many) "tail" composed of a long string of adenosine. Chemical methods exist to find this adenosine tail, serving as the basis for the chemical methods to isolate mRNA.

Remember that the mRNA is not exactly the genetic code. In fact, it is the inverse the code because it contains the complementary base for each organic base on the DNA molecule. For example, as you remember, where the DNA has adenosine, the mRNA will have Uracil. Happily, there are naturally occurring enzymes that will translate the information on mRNA and go backwards to DNA. These are called *reverse transcriptases*. The mRNA is used as a template by the reverse transcriptase to read out a complementary strand of DNA. When the messenger RNA is removed, the single-strand DNA produced by the reverse transcriptase combines appropriately with the complementary nucleotides, through the action of DNA polymerase. The result is the production of a new, a double-stranded molecule known as *complementary DNA* (cDNA), as shown in Figure 7-1. This is step 2 of the process presented in Figure 7-8.

INSERTING THE HUMAN GENE INTO BACTERIAL DNA

You really want to insert your insulin gene into bacteria because the bacteria will produce the human protein in a form much less likely to be rejected by the recipient patient's immune system than animal insulin would be. Also, the cost of harvesting the protein from your bacterial colony is much less than the cost of maintaining the large animals that formerly were used to obtain insulin. The challenges in inserting a human gene into a bacterial DNA include the following:

- First, getting the DNA into the cell is difficult. You need a *vector* that will transport the DNA into the cell.
- Second, once inside the cell, the bacteria will recognize human DNA as foreign, primarily based on the fact that the bacterial DNA is circular and yours is linear. The bacteria will destroy your DNA. You need to insert the DNA in a circular form.

Most bacteria have a single, circular chromosome. Fortunately, they also contain small bits of ancillary DNA in structures called *plasmids*. Plasmids, as in Figure 7-2, are small ring structures that exist independently of the major chromosome in bacteria. Typically, they carry less than 5 percent of the cellular information and don't contain information that is essential for cell survival. They tend to carry a few genes that pertain to survival in abnormal environments. All this suits your purposes just fine because the plasmids can serve as your vector.

You will cut the plasmid open with *restriction enzymes*. Along with the discovery of reverse transcriptase, the discovery of these enzymes that occur naturally in bacteria was crucial in the development of genetic engineering. The first restriction enzyme was isolated from bacteria and characterized by H. O. Smith, K. Wilcox, and J. J. Kelly in 1968. These naturally occurring enzymes cut apart DNA molecules at specific base sequences. This is step 4 of the process presented in Figure 7-8.

Why would bacteria contain enzymes that cut DNA apart? Possessing these enzymes is beneficial to the bacteria because they protect against *bacteriophages*,

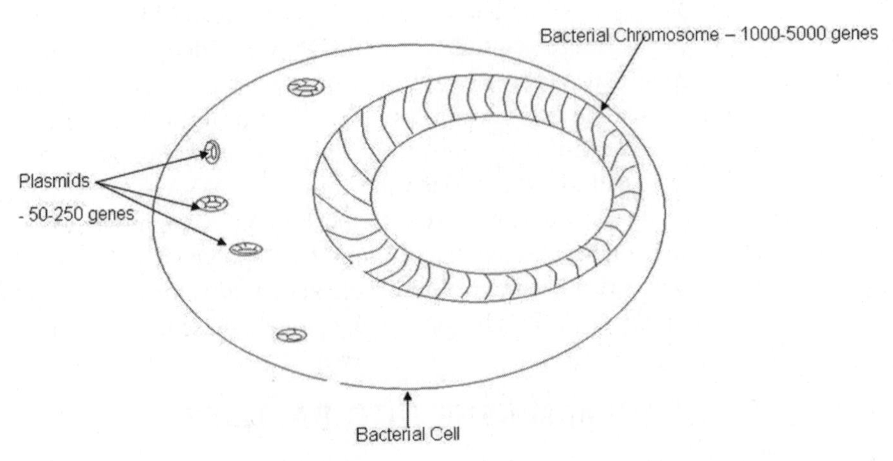

Figure 7-2 Plasmids

or viruses that infect bacteria. When the viral DNA injects itself into the bacteria, this DNA is attacked by the restriction enzymes and the viral DNA is destroyed. The bacteria's own DNA is protected by the presence of unique methyl groups on bacterial DNA base pairs. These methyl groups prevent the action of the restriction enzymes, of which over 900 have been identified. Each enzyme has its specific site of action, and there are over 200 unique DNA sequences that are recognized by restriction enzymes. Restriction enzymes are specific to sequences of 10 base pairs in the DNA. In the example in Figure 7-3, the restriction enzymes cut the DNA between base pairs G-C and T-A.

You can snip your plasmid anywhere you want by selecting the restriction enzyme that cuts at the base sequence you select. Furthermore, you know that if you cut a DNA with restriction enzymes, you are blessed with "sticky ends" on each end of the DNA. This is because the restriction enzyme makes a jagged cut; the ends are not literally sticky but they hold short segments of single-stranded DNA. These short segments will readily combine with new DNA that contains the complementary base sequences. In Figure 7-3, the sticky ends consist of the segments G-A-T-T on one side and C-T-A-A on the other. These ends will "stick" to segments C-T-A-A on the first side and G-A-T-T on the second. Additional base pairs can be added between these sticky ends. You can isolate plasmids from bacteria and use restriction enzymes to cut a hole in the bacterial plasmid. Then you may stick on the gene that you want.

You have to generate sticky ends on your gene, too, by using the appropriate restriction enzyme. By choosing the restriction enzyme, you select the nucleotides

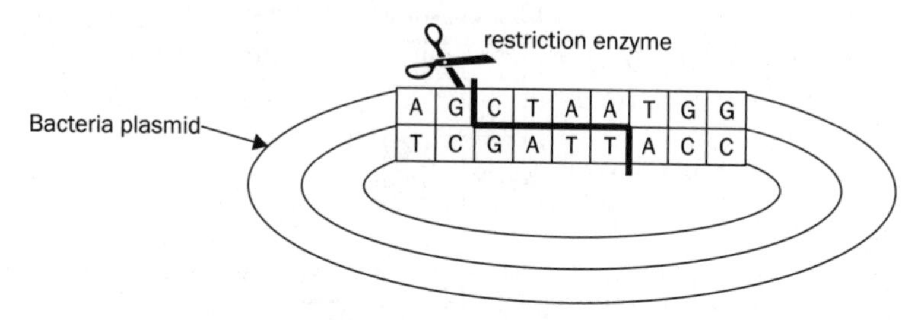

a. Restriction enzymes make a jagged cut across the DNA

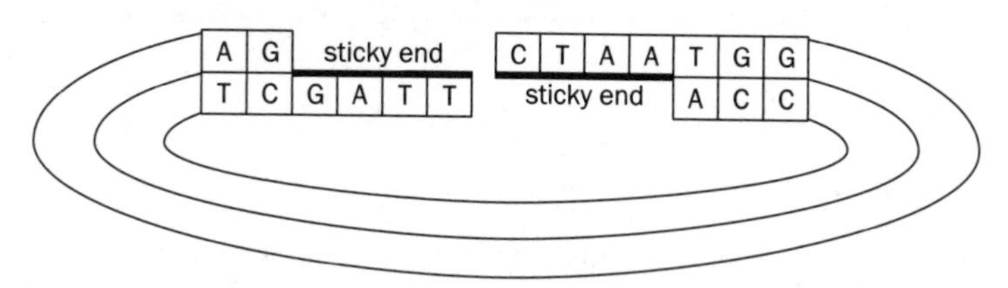

b. Exposed single strands are available to mate with matching base—i.e. are "sticky"

Figure 7-3 Restriction enzymes

that will be on the overhanging (sticky) end of the broken DNA. But what if the gene that you want doesn't have a comparable sticky end that will attach to the broken plasmid? The biotechnology industry has anticipated your need and developed little DNA snips that contain all the sequences you would want to complement the sticky ends. These sequences can be added to your gene using an enzyme called *DNA ligase*. This is step 3 in Figure 7-8. You can now add your gene to the plasmid and create man-made, *recombinant DNA,* as shown in Figure 7-4. This is step 5 in Figure 7-8. Remember that you started with mRNA from islet cells because these cells produce large quantities of insulin and you are trying to obtain a gene for insulin. From the mRNA you started with, a large percentage (but not all) of the genes you create will code for the insulin you want. Some of the plasmids will take up the insulin gene. However, islet cells produce other proteins besides insulin and some of the genes you have isolated will be for these other proteins. You have no way of selecting only insulin genes at this point, so after incubation with plasmids, some of the plasmids will take up a different gene. And some of the plasmids will not take up new genetic material. In summary, some plasmids contain a gene for insulin, some plasmids contain a gene for other proteins, and some plasmids contain no new genes; you cannot distinguish these plasmids. Now you add all the plasmids, only some of which have the gene of interest, into a population of bacteria. The bacteria are induced to allow plasmids to enter by treatment with a solution of calcium chloride, a process known as *transformation*. This is step 6 in Figure 7-8.

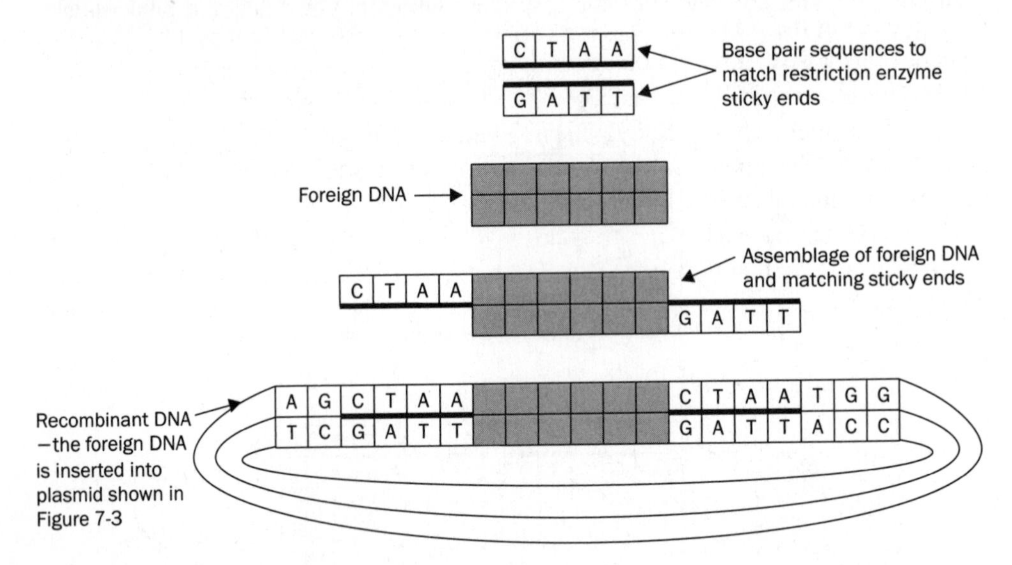

Figure 7-4 Inserting DNA into a bacterial plasmid to create recombinant DNA

SELECTING ALTERED BACTERIA

Now you have a diverse population of bacteria—some have taken in plasmids with the gene for insulin, some have taken in plasmids with other genes, some have taken in plasmids with no new genetic material, and still other have not taken in any plasmids. You need the subpopulation of bacteria that have taken up the right plasmid with the right gene inserted.

You can eliminate bacteria that have not taken up any plasmid by culturing them in an environment where they cannot survive without the native gene supplied by the plasmid. Remember that plasmids tend to house genes that enable survival in abnormal environments. Typical genes to be found on plasmids are genes for resistance to antibiotics or metals. You can choose to use ampicillin-sensitive bacteria and to insert plasmids that contain a bacterial gene that provides ampicillin resistance. If you culture the bacteria after transformation in the presence of ampicillin, only those bacteria that have taken up the plasmid will survive. You therefore eliminate bacteria that have not taken in plasmids, as shown in Figure 7-5. This is step 7 in Figure 7-8.

Now you have eliminated all bacteria that have not taken up a plasmid. The next step is to eliminate bacteria that have incorporated plasmids with no new genetic material. Your goal is to find the bacteria that have taken up the plasmid with the correct gene insert. Remember you can select the site on the plasmid where the gene is placed because you know the gene sequence and you have selected the restriction enzyme to cut the plasmid at a specific spot. You can choose to interrupt a code for a gene that you can detect in culture. One system commonly used is beta-galactosidase. This enzyme will generate a blue color if provided with a substance known as X-gal (5-bromo-4-chloro-indoylgalactoside). If you engineer the system such that the plasmids take up the gene insert in the middle of the beta-galactosidase gene, you will no longer have a functional gene for beta-galactosidase. You can observe the bacteria that have taken up the gene insert because these bacteria will not generate a blue color in the X-gal medium. The beta-galactoside gene is called a *reporter gene* because it reports on the condition of the plasmid DNA. You culture your

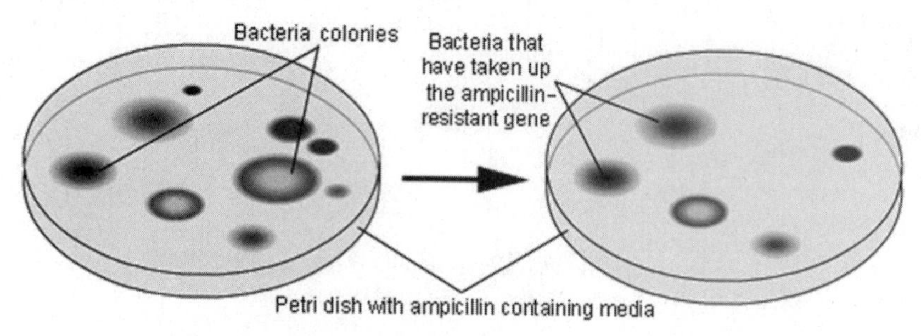

Figure 7-5 Eliminating bacteria with no plasmid insert

bacteria in agar containing X-gal. Colonies with the beta-galactosidase gene will turn the agar blue; you want the other ones, as shown in Figure 7-6. This is step 8 in Figure 7-8.

You have narrowed your bacteria population down to the bacteria containing plasmids with new genetic material. Your original goal was to obtain a population of bacteria that would produce human insulin. Now you need to find the bacteria that contain the gene for insulin. Not all of the new genetic material codes for insulin. Remember that you collected mRNA from pancreas cells. Most of the mRNA will code for the production of insulin, but some of it will call for other proteins.

You need to be able to visualize the insulin that may be produced. You can flag the insulin-producing cells by the use of monoclonal antibodies. Monoclonal antibodies were described in detail in Chapter 5. For our purposes, monoclonal antibodies are antibodies that are specific to a given protein or part of a protein. The ability to produce monoclonal antibodies was a very nice advance in diagnostic tools for biomedical research because such antibodies are very specific and because you can attach labels to the antibodies, labels that don't interfere with their ability to bind antigen. You can produce antibodies against insulin and you can label these antibodies in several ways. Fluorescent dyes can be attached to the antibodies and be made to fluoresce when they bind to cells containing insulin. Another possible label is radioactive molecules that are taken up by the antibody. An example is radioactive hydrogen, known as tritium. The tritium molecules are taken up by antibodies and replace nonradioactive hydrogen in the molecular structure. The presence of the tritium can be visualized by placing a photo-film over the material in question. The radioactive emissions from the tritium will expose the film, just like light does.

The monoclonal antibodies against insulin will need access to the interior of the cells in order to bind the insulin. You can cause the cells to burst (known as *lysing* the cells), thereby releasing their contents. The reason you need to lyse the cells is that the antibodies cannot pass the cell membrane to enter the cell interior.

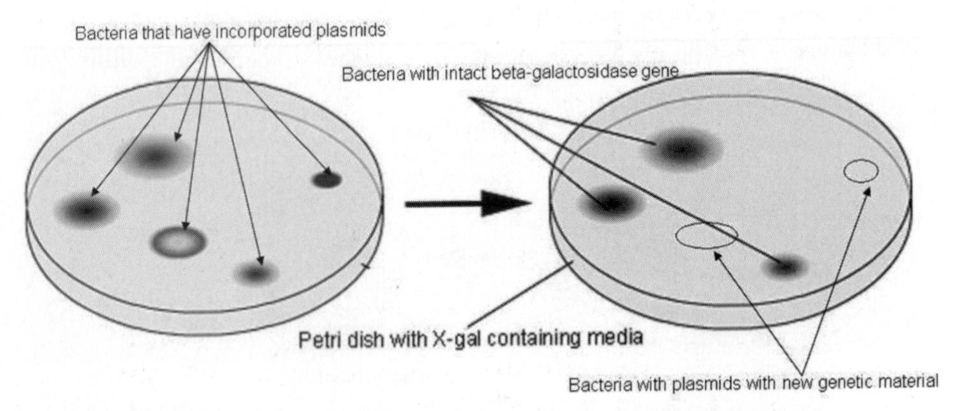

Figure 7-6 Selecting bacteria with the correct plasmid

Of course, you don't want to kill all the bacteria that you have gone to so much trouble to produce, so you transfer some of each colony to new agar plates. The way to do this is to place a piece of cloth over the colonies of bacteria. The cloth picks up a few members of each colony oriented exactly like the colonies are on the plate. If you press this cloth against a new agar plate, the agar will pick some of the bacteria on the cloth and new colonies will grow as an exact replica of the original. You can lyse the new cells without disrupting their orientation on the plate using a number of standard techniques. You incubate these lysed cells in an assay with the monoclonal antibodies to see which colonies contain insulin. You find the antibodies that have attached to the target by looking for either your fluorescent or your radioactive tag. Once you have isolated a colony of bacteria that binds the antibody and therefore contains the protein you want, you can culture this particular colony of bacteria on the original place to produce the protein, as shown in Figure 7-7. This is step 9 in Figure 7-8.

You may wonder how you make sure that your host cell is motivated to produce insulin. After all, every cell in the human body has the equipment to produce insulin, but only the cells of the islets of Langerhan read out, or *transcribe*, the gene and make the protein. Just because the bacteria contain the gene for insulin doesn't mean that they will produce insulin. Once again, you take advantage of the fact that you can place the gene anywhere you want on the plasmid by your selection of restriction enzymes, assuming you know the gene sequence on the plasmid.

Remember from Chapter 4 that genes are transcribed by an enzyme called RNA polymerase. The RNA polymerase recognizes a binding site on the DNA that says, "start." You can place your gene forward of a site where gene transcription begins, the start site, even though naturally this start site calls for another protein. Remember that the RNA polymerase, after binding at the start site, chunks out mRNA until

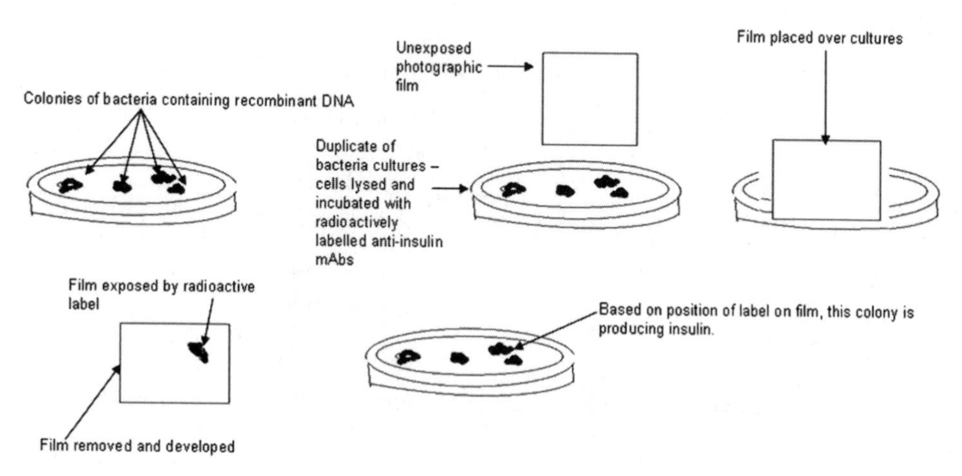

Figure 7-7 Selecting bacteria expressing the correct gene

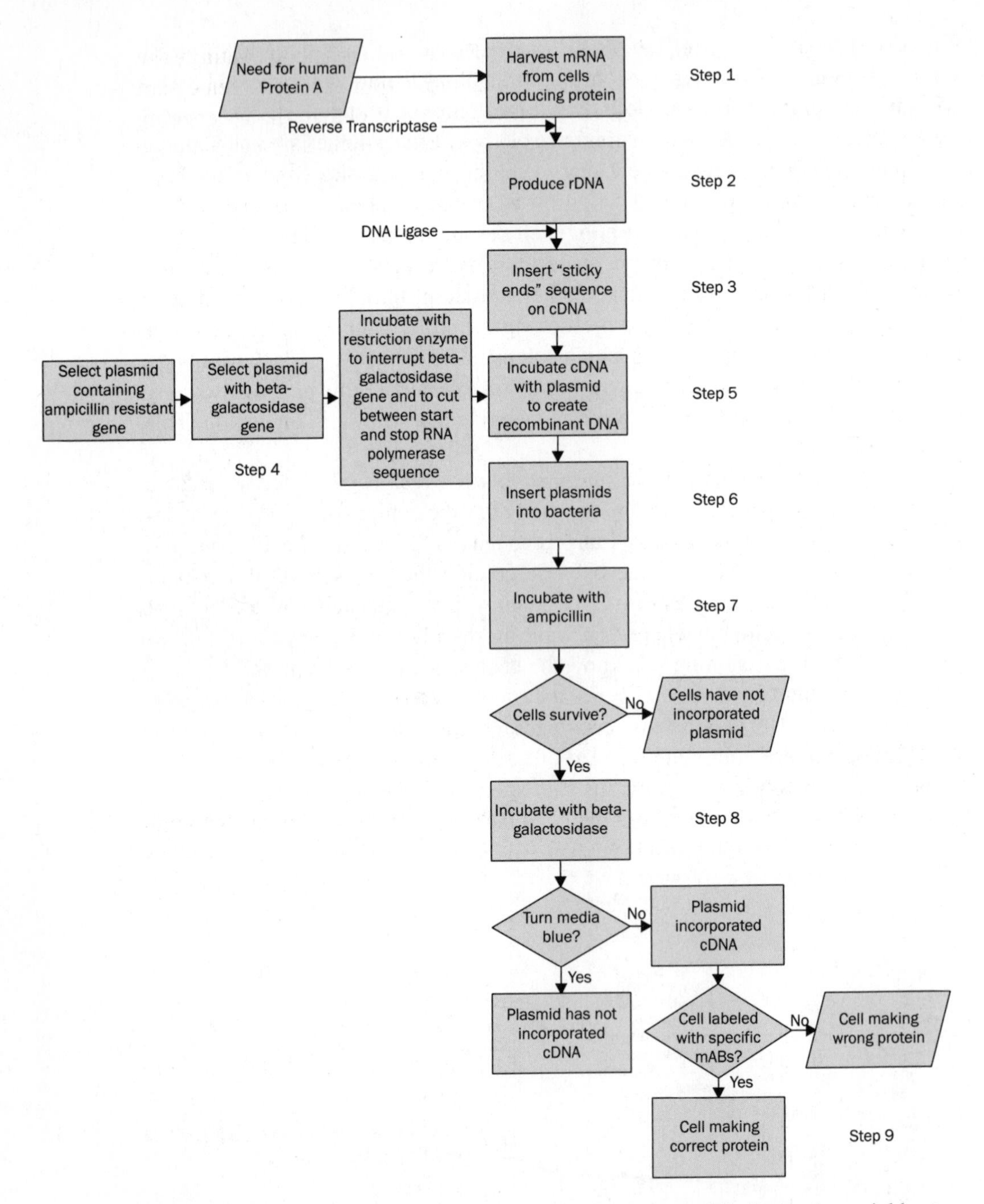

Figure 7-8 Flowchart of the entire process—producing a human protein from a bacterial host

it hits a stop signal—utterly indifferent to the protein sequence it is creating. The mRNA goes on to make your protein. So, if you sneak your gene between a start site and a stop sequence, the cell's RNA polymerase will oblige you by creating mRNA for your protein. Plasmids that can be engineered this way are called *expression vectors* because they ensure that your protein will be produced, i.e., that the gene will be expressed.

SELECTING THE CORRECT HOST

Normally, cells have processed to break down unneeded or foreign proteins. You need to select a host that will not attack and destroy the insulin in the cell before it can be secreted. To find host cells that will tolerate the foreign protein, you can develop subpopulations of your cell by growing these cells under different conditions. You can select cells that are deficient in the proteolytic enzymes necessary to degrade the protein, and you may need to develop special strains of bacteria. For some proteins, it may be necessary to use a eukaryotic cell, for example, if the protein is toxic to bacteria cells. If the goal is to research gene expression, then eukaryotic cells are by far the better choice. The most popular single-cell eukaryotic organisms for DNA research is baker's yeast (*Sacharomyces cerevi*). Eukaryotic cells are more difficult to maintain in culture and require a different type of vector (see the discussion that follows).

VECTORS

By definition, *vectors* carry recombinant DNA into an intact cell. For insertion of small amounts of DNA into prokaryotic cells, plasmids are routinely used and the insertion is accomplished by treatment with calcium chloride, as described earlier. This system works very well for small amounts of genetic material. However, for larger genes, other vectors must be considered.

Viruses are an attractive option. Viruses were not included in the discussion of cell types because viruses are not cells. They are little snips of DNA, enclosed by a capsule. Their only function is to replicate themselves. They do not produce proteins or perform any other function typical of a real cell. They reproduce by commandeering the cell's DNA replication mechanisms. They inject their DNA into the cell. Usually, the viral DNA invades the host's DNA, and when the host DNA divides, the virus is also replicated. Other types of viruses, called *retroviruses*, are composed RNA and produce DNA when they invade the cell. They may not invade the host DNA. Unhappily, virus particles, once replicated, then seek a way out of their host, usually by lysing (breaking apart) the cell. If you could sneak your DNA into a virus, the virus would do what it always does and invade a host cell DNA. The host cell DNA would read out for your protein. You do, however, need to prevent the virus that carried your DNA into the cell from lysing the cell.

Viruses that prey upon bacteria are called *bacteriophages* or simply *phages*. The bacteriophages are used as vectors if the gene of interest is to be transported into a bacterium and is too large to be carried on a plasmid. They have the disadvantage of inserting their genetic material rather randomly into the bacterial cell. Remember, with a plasmid, you, the engineer, determine where the new gene will be placed. Also, again, you, the bioengineer, need to manipulate the cellular environment so that the virus does not induce lysis of the cell. Nonetheless, phages are widely used to provide new genetic material to bacterial hosts.

Viral vectors are also being used in experimental genetic therapy for human beings. Human viruses, just like bacterial phages, insert new DNA into human chromosomes. You can introduce a gene for a new protein into a human cell by this method. The advances in providing human beings new genetic material to treat diseases with a genetic basis will be discussed in detail in Chapter 10.

The placement of the genetic material is of special importance when the genetic experiment is conducted within a human patient. You might inadvertently interrupt a gene that produces an important protein, or you might create a cancer-causing gene. Researchers became concerned that the insertion of new genetic material into certain regions of human DNA could lead to cancer when two very young victims of SCIDS (Severe Combined Immunodeficiency Syndrome) developed leukemia after gene therapy. The SCIDS victims were treated using a vector made from a retrovirus. These viruses may insert themselves into the human DNA to be effective. The vector used in experimental trials on victims of hemophilia and cystic fibrosis is derived from an adeno-associated virus. This virus is less likely to insert itself into the host DNA and can be effective as a plasmid. However, recent studies show that this vector also will insert into the host DNA and, for unknown reasons, is more likely to insert itself into genes than into the noncoding portions of the DNA. Proponents of gene therapy point out that there is limited evidence that infection with naturally-occurring inserting viruses predisposes the host to cancer and that the recipients of the gene therapy are subject to life-threatening diseases with limited other therapy options. However, researches are seeking vectors that insert at known sites or don't insert themselves at all into existing DNA. Current options include the use of endonucleases to guide the insertion into the host DNA, or of artificial chromosomes that remain distinct from the host DNA (see the next section).

Zinc endonucleases are naturally occurring enzymes that intersect DNA at predictable points. Endonucleases can be either two- or three-fingered. The fingers are the sites where the endonuclease binds to the "doomed" DNA. If there are two fingers, then the endonuclease binds at two points and the binding site for each finger is a specific DNA sequence. If the molecule is a three-fingered endonuclease, then

there are three binding sites that must carry a specific sequence. The three-fingered endonuclease is a more restrictive enzyme. In experimental settings where two-fingered endonucleases are used to influence the location of the insertion of the new genetic material, the insertion has occurred in the correct area in only a small percent of the insertions. Recent studies have indicated that the use of the three-fingered endonucleases greatly improves the percentage of gene insertions that occur in the correct place.

ARTIFICIAL CHROMOSOMES

Plasmids are too small to carry much of the genetic material we would like to use and, as just discussed, viral vectors are difficult to control. The problems with vectors could be solved if we could build our own chromosomes with whatever genetic material we desired. These chromosomes would coexist with the natural genetic material of the cell much like the DNA of mitochondria does.

Researches have discovered that double-stranded DNA sequences that are provided with telomeres at the ends, a centromere, and replication sites for the binding of DNA polymerase will behave like chromosomes. This discovery has allowed the development of *artificial chromosomes*. As you may remember, telomeres consist of dense sections of DNA and protein, and are located at the ends of chromosomes to protect them from damage. *Centromeres* are specialized regions of DNA, essential for the proper control of chromosome distribution during cell division. The centromere and DNA polymerase binding sites allow the artificial chromosome to be duplicated during cell division, along with the host DNA.

The most widely used artificial chromosomes are *bacterial artificial chromosomes* (BACs), developed in 1992. These are based on natural bacterial structures to provide the bacterial telomeres and centromere. The most popular basis for the natural structures that you need on your BAC is a plasmid called the f (or fertility) plasmid from the organism *Escherichia coli.* The BAC system allows a much larger segment of DNA to be inserted in the cell than would be possible with traditional recombinant DNA techniques. Up to 300,000 base pairs can be inserted in a single BAC. The BAC can be transferred into the cell by *electroporation*, where a series of electrical pulses causes pores to form in the cell membranes.

BACs are known to be highly stable within the bacterial genome. This stability explains why BACs are so useful for sequencing genetic material. As the bacteria grow and divide, the BAC is replicated, and the result is a colony of genetically identical cells, each containing a copy of the target DNA. The target DNA has thus been amplified and can be subsequently isolated from the rest of the DNA inside the cells. In the Human Genome Project, many fragments of the human genome were incorporated into BACs where they were cloned prior to sequencing.

Artificial chromosomes based on the chromosome structure of organisms other than bacteria have been developed. The *yeast artificial chromosome* (YAC) was first developed in 1987 by David Burke. The YAC contains the telomere, centromere, and origin of replication sequence elements of yeasts. The engineered YAC is put into a yeast cell by a chemical means that induce the cell to take up the genetic material. The synthetic chromosome remains independent of the rest of the genetic material within the host cell and functions essentially as an accessory chromosome.

The advantage of the YAC over the BAC is that the former can carry a much larger segment of DNA, up to a million nucleotide pairs. Also yeast is much better in tolerating foreign DNA that contains highly repetitive sequences (like human DNA). The disadvantage is that YACs are much less stable than BACs. Harvesting the cloned DNA is more difficult. Therefore, YACs are more useful in the early stages of genome research.

In 1997, researchers at the School of Medicine and Athersys, Inc., created the first *human artificial chromosomes*. (Huntington F. Willard, *Nature Genetics,* April, 1997). This development was *not* accomplished by building human artificial chromosomes and introducing them into a human cell. Human centromeres consist of large segments of highly repetitive DNA, called *alpha satellite DNA*. The researchers synthesized the alpha satellite DNA and then introduced the resulting centromeric material into human cells, together with telomeres and genomic DNA. Inside the cells, the independent elements assembled to form miniature chromosomes, called *synthetic microchromosomes*. The new microchromosomes demonstrated normal gene expression through repeated rounds of the cell cycle, that is, they produced new proteins. The clinical use of artificial chromosomes would prevent the risks introduced by viral vectors, including potential chromosomal damage or interruption of normal gene expression when the new material is inserted into the existing genome. However, devising a regimen where human patients can undergo the processes necessary to create artificial microchromosomes *in situ* (within the living organism), has proven to be challenging.

Deciphering DNA

We have talked about ways to insert genetic material into a genome and about creating recombinant DNA, but we haven't discussed how a genome is deciphered. The Human Genome Project determined the DNA sequence for the entire human genome. This development will shake the medical community when the clinical outcomes finally hit the street, but how was this miraculous accomplishment done? The techniques are surprisingly simple.

POLYMERASE CHAIN REACTION

Many genetic engineering applications depend on our ability to make many copies of a given DNA fragment. In fact, the development of the *polymerase chain reaction* (PCR) is arguably the key technology that has allowed the biotechnology revolution. A very small sample of DNA can be magnified many times to allow analysis. This technology has been instrumental in the successful completion of the Human Genome Project as well as other applications, such as DNA analysis to identify perpetrators of crimes or to indemnify suspects of crimes.

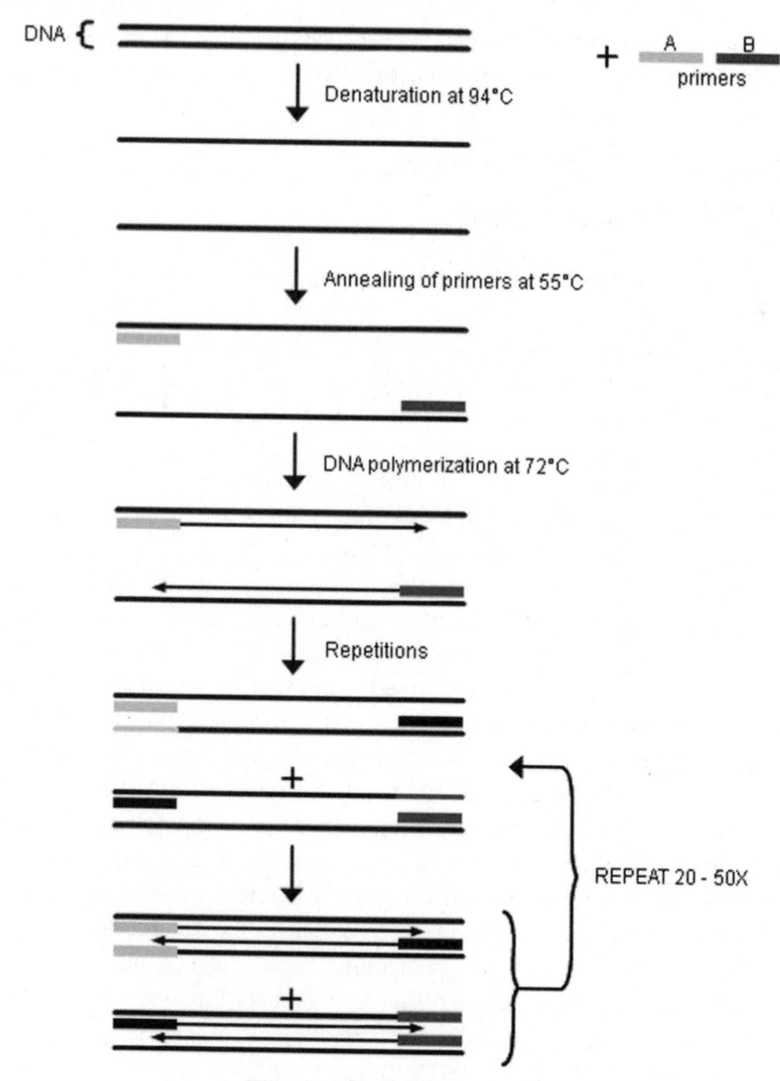

Figure 7-9 PCR techniques

There are three basic steps in PCR, as shown in Figure 7-9. First, the target genetic material is heated to 90–96°C. This denatures the material, causing it to unwind and separate. The individual strands are processed separately. In the second step, short segments of complementary bases, called *primers*, are attached to the end of the now single-stranded DNA. In the third step, the polymerase reads a template strand and matches it with complementary nucleotides very quickly. The polymerase will make two new DNA double strands out of the original one. With this technology, a small segment of DNA can be used to generate enumerable copies.

A major component in the successful implementation of PCR is a form of polymerase that is tolerant of high temperatures and is functional in the rapidly changing thermal environment of the automated PCR process. This polymerase is affectionately called *Taq*, short for *Thermus aquaticus*. Taq dwells naturally in hot springs.

DNA PROBES, HYBRIDIZATION, AND THE SOUTHERN BLOT

A fundamental instrument in revealing the base sequences on an existing DNA strand is called a *nuclear* or *DNA probe*. The phenomena behind DNA, or nuclear, probes is the inclination of single-stranded DNA to combine with strands of DNA containing base sequences that are complementary to the base sequences of the original strand. If you are looking for a base sequence of A-T-C-G-G, you would use a probe containing the sequence T-A-G-C-C. The resulting double strand is called a *hybrid* because it is a combination of the natural DNA and your "unnatural" probe. The hybridization step consists of simply mixing the single-strand probes with the target DNA (the combined molecules).

The DNA has to be denatured first. *Denaturation* of the DNA is obtained by heating the DNA, which separates the two strands and allows access of the single-strand probes to their respective complementary single target strand.

The use of a nuclear probe is often coupled with a detection technique known as the *Southern Blot*. The Southern Blot technique was named after its developer, Edward Southern, and is used to analyze DNA fragments. Human nature being what it is, we now have a Northern Blot technique that is applied toward RNA. Western Blots are devoted to proteins. Future generations will undoubtedly be mystified by these names.

To apply the Southern Blot technique, you must first digest your DNA segment of interest into small fragments. You use one of our trusty restriction enzymes to accomplish this cutting or digestion. You will be able to determine the DNA sequences at the sites where incision occurs. Then you run the resulting DNA fragments through gel electrophoresis. *Electrophoresis* is an analytical technique to separate molecules based on their size and/or their electrical charge. The compounds of interest are induced to migrate through a gel, such as agarose, by applying an

Figure 7-10 Gel electrophoresis

electrical charge across the gel. The speed at which they move is dependent upon the size of their electrical charge, because they will respond to the electrical field according to how strongly they are charged. Their speed also depends on their size because the gel is actually a matrix of interlocking fibers. Smaller compounds find it easier to navigate through the matrix than do larger compounds. The various fragments that you generate are placed at one end of the gel in a well. Each fragment gets its own well. Then the electrical charge is applied. See the result in Figure 7-10.

You don't see anything? You have to do something to visualize your DNA fragments. Soak the gel in ethidium bromide (EtBr), which binds to DNA and is fluorescent. When you expose the DNA to the correct wavelength of ultraviolet light, you can visualize your DNA fragments, as shown in Figure 7-11. There!

But what about your probe? The ethidium bromide showed all the fragments, but you have no way of knowing which fragment bound your probe. Hopefully, you thought ahead and provided your probe with some kind of label, either radioactive or fluorescent. One way to detect target molecules is with a system of coupled antibodies and fluorochromes, a method known as *fluorescent in-site hybridization* (FISH). The probes can be synthesized with incorporated fluorescent molecules or molecules that can be recognized with fluorescent antibodies so that the direct visualization of the probes is possible.

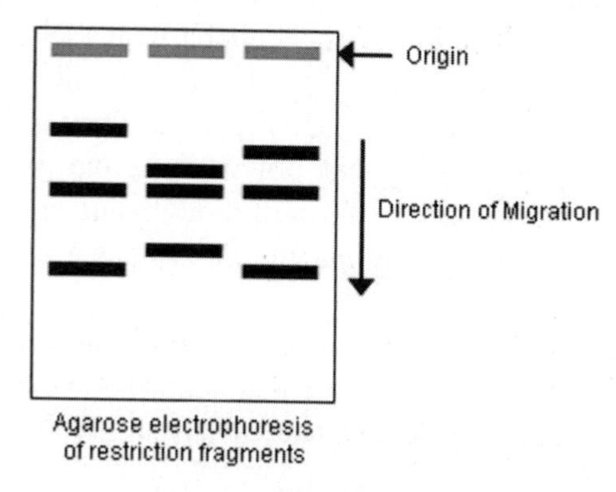

Figure 7-11 Visualization with ethidium bromide

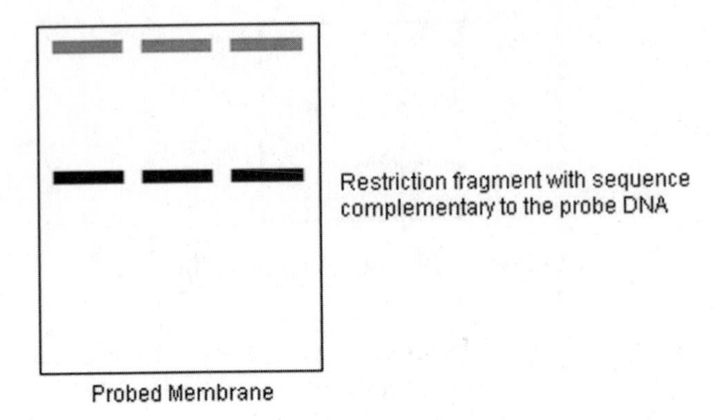

Figure 7-12 Location of DNA probe

In either case, you will dry the gel out on a filter system to provide a solid matrix for the analysis. If you have labeled your DNA fragments with a radioactive atom such as tritium, you will lay a photosensitive material over the matrix (film) and the radioactive material will expose this material just like light does. If you have labeled your DNA fragment with a fluorescent label, then you will expose the matrix to UV light of the appropriate wavelength. Obviously, you will need a label that requires a different wavelength of light than does EtBr. In any case, you should see something like Figure 7-12.

From this analysis, you learn the location of the DNA sequences specific for your restriction enzymes and the location of the base sequences complementary to your probe within the DNA fragment.

SANGER TECHNIQUE

The *Sanger technique* has been used since its development in 1977 for DNA sequencing. It is called the "chain terminator method" because it is a very clever method of determining the location of a specific base on a DNA fragment, based on where synthesis of a new DNA chain stops. The methodology depends on the fact that (a) synthesis of a double-stranded DNA segment from a single strand of DNA will be initiated in the presence of DNA polymerase, and (b) DNA synthesis will stop if the incorporated base is in the form of dideoxynucleotide instead of deoxynucleotide. The dideoxy form of the nucleotide is missing a hydroxyl group at a critical point. See Figure 7-13.

So, if you provide a batch of synthesizing DNA molecule with, for example, dideoxynucleoadenoside (ddATP) in a mixture that also contains deoxynucleoadenoside (dATP), as well as the other three deoxynucleotides, the synthesize of the double chain

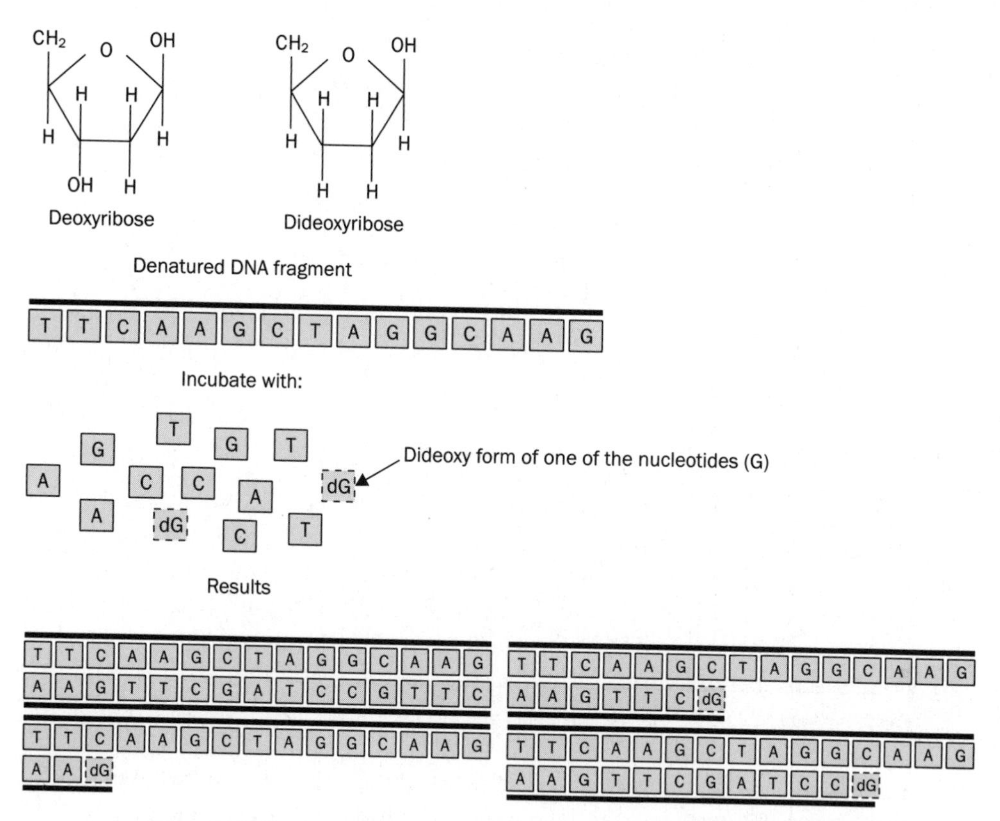

Figure 7-13 Use of dideoxynucleotide to terminate DNA synthesis

will stop when the ddATP molecule is incorporated instead of the dATP molecule. By the laws of probability, some of the synthesizing DNA will be stopped at every point that adenosine is required.

The Sanger technique uses the dideoxynucleotide for all four of the required nucleotides. There are four batches of reagents, one devoted to each nucleotide. The same single-stranded DNA molecule is incubated in each batch with one of the nucleotides provided in the dideoxy form as well as its normal deoxy form. Among the four batches, synthesis has been arrested at every site in the DNA fragment. You keep the batches separate and run gel electrophoresis on all four batches. Because the length of the migration in the electrophoresis field depends on the size of the molecule, the DNA fragments should distribute themselves in linear fashion according to size. To use the ddATP example, there will be some DNA fragments truncated at each adenosine site. Some of these fragments will be small, where the adenosine appears early in the chain, and some will be long, where the adenosine appears later in the DNA chain sequence. You can tell the location of the adenosines by the length of these various

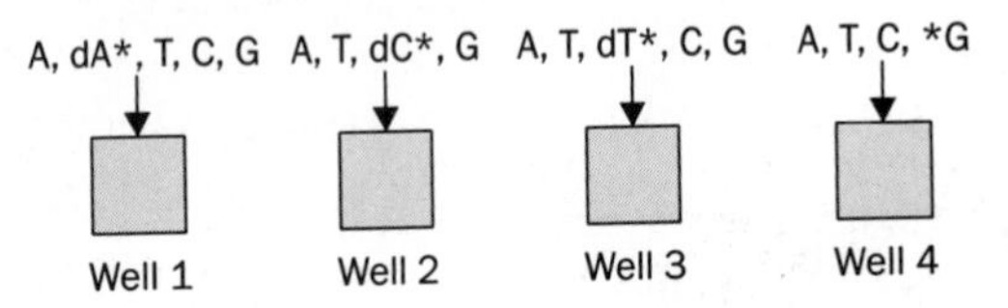

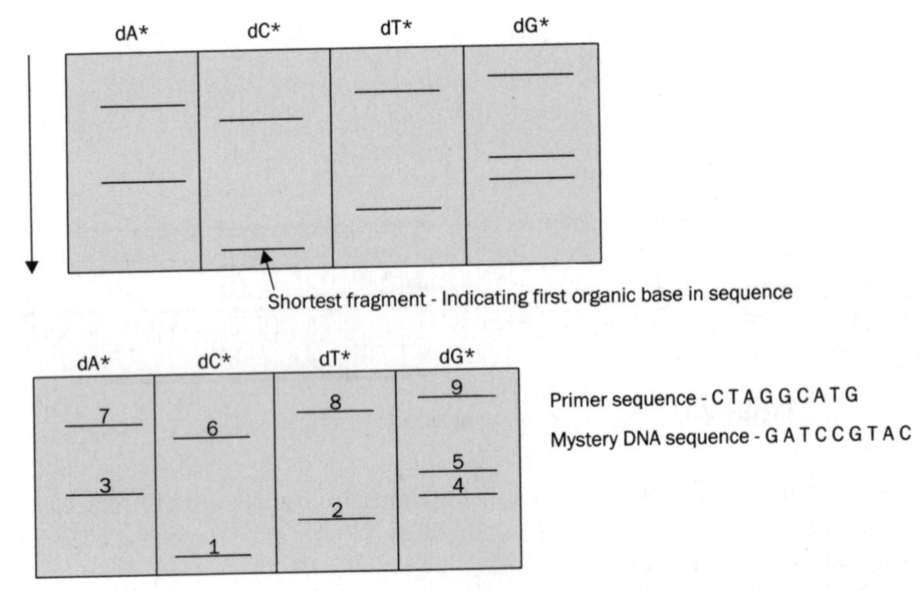

Figure 7-14 Determining DNA sequence using the Sanger technique

fragments. See Figure 7-14 to understand how, when you compare the four runs of gel electrophoresis, you can tell which base has been incorporate at every site in the DNA fragment.

MICROARRAY TECHNOLOGY

Among the commercial opportunities presented by the science of bioengineering is the development and marketing of DNA microchips or *genetic microarrays* for numerous targets. These are large matrices of gene fragments. You can buy microarrays created from the DNA material of all types of organisms and of human cells.

These microarrays are fundamental in research to determine which genes are expressed within given cells. They are very important in determining the genetic basis for the difference between normal and abnormal cell function.

To illustrate how microarrays are used, consider the researcher who is studying the difference in cell function between pancreatic cells of normal individuals and individuals suffering from diabetes. He or she will harvest mRNA from both types of cells. The mRNA tells the researcher which genes are expressed, in other words, which genes are calling for the production of proteins. The researcher will use this mRNA to re-create the DNA from which it came by use of reverse transcriptase. Remember that this new DNA is called cDNA. The cDNA from the normal cell will be identified with one fluorescent label, say green for the sake of illustration. The cDNA from the abnormal cell will be identified with a different fluorescent label, say blue. The researcher will obtain a DNA chip with denatured DNA fragments from the genome of pancreatic cells. The DNA fragments are arranged in an array on a microscope slide. The location and identity of each fragment is known. In the location two down from the top and three over from the left, the research might know, for example, that a gene fragment with the sequence A-T-C-G-C is present. The microarray is incubated with the cDNA from both the normal and abnormal cells. The cDNA will fix on the matrix by binding with the denatured DNA fragments that contain the matching (or mirrored) base pairs. By scanning the array with first a green and then a blue laser, the researcher will know which genes are expressed by both cells, which genes are expressed by only the normal cell, and which genes are expressed by only the abnormal cell. The microarray is like a switchboard; the operator can tell at a glance which circuits are turned on. Consider the example in Figure 7-15 of mutant bacteria that can digest polychlorophenols (PCBs) versus normal bacteria.

REPORTER GENE TECHNOLOGY

Reporter genes are used when you are investigating a gene and have no easy way to determine if the gene is being expressed. Reporter genes are active when the gene under study is active; systems that are under study can include viral DNA and mutant genes. Some reporter genes make an enzyme that is easy to detect, such as the gene that codes for beta-galactosidase. Beta-galactosidase causes a color change in the culture media, as described earlier. Another popular reporter gene is the gene than encodes for a green fluorescent protein. The reporter is tied to the event under study, for example, the effect of human growth hormone on a cell. The reporter could be fused to an HGF-sensitive promoter and would glow when the HGF was influencing cell function. The results are spectacular. The cells literally turn an Incredible-Hulk green.

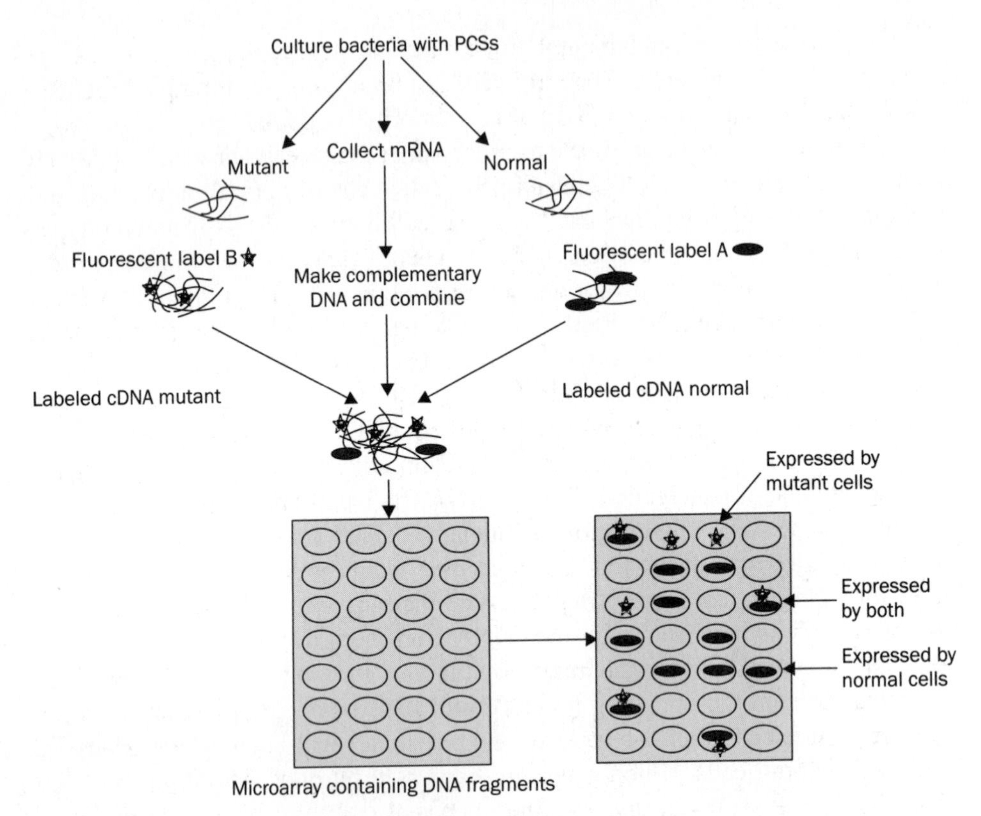

Figure 7-15 Example of a microarray

Another use of reporter genes is to track the movement of proteins. In a really beautiful study of the normal and abnormal metabolism of a gonadotropin-releasing hormone (GnRH) receptor, researches fused a gene for the green fluorescent protein to the gene for the receptor. They were thereby able to see where the protein was within the cell because the protein was fluorescent green. In this wonderful study, if the protein was not folded properly, it accumulated in the lumen of the endoplasmic reticulum. Properly folded proteins were observed to insert properly into the cell membrane and appeared at the cell surface. (P. Michael Conn and Jo Ann Janovick, "A New Understanding of Protein Mutation Unfolds", *American Scientist*, Vol 93, no. 4, pp. 314–321).

To the delight of the popular press, the gene for the green fluorescent protein can cause entire organisms to glow; you may see pictures of green fluorescent mice. A typical and very flashy reporter is luciferase, made from firefly DNA. See Figure 7-16.

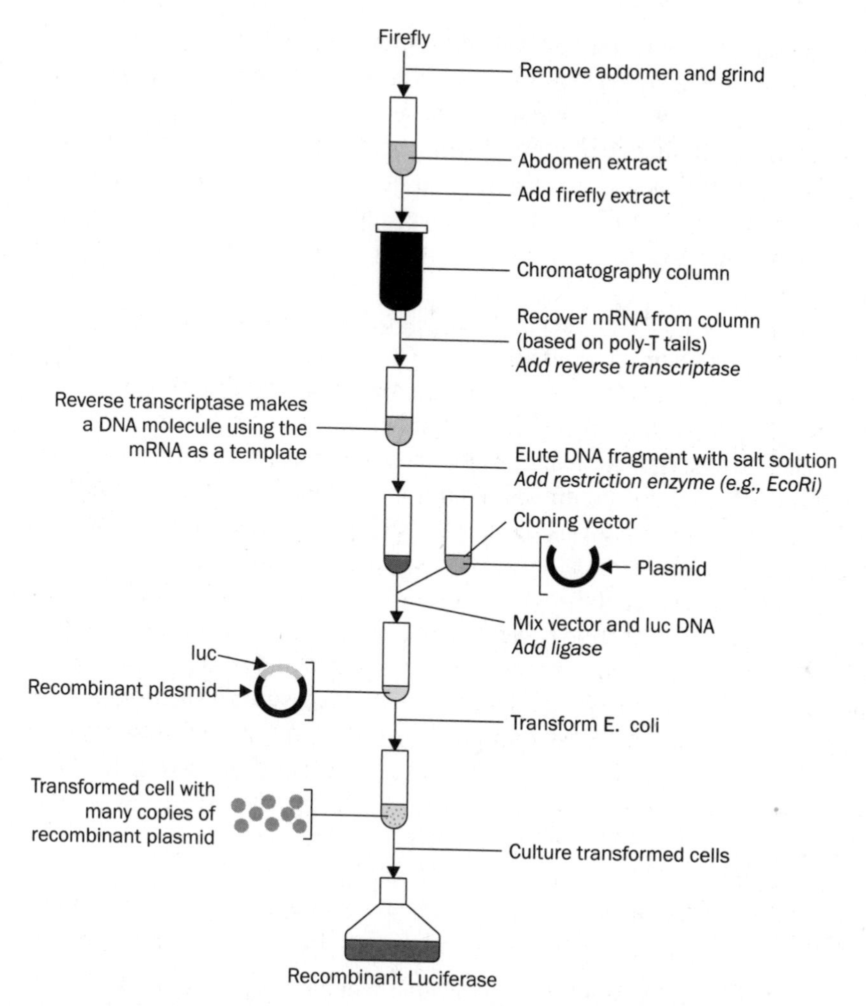

Figure 7-16 Making recombinant luciferase (Based on Figure 8-12 of *Schaum's Outlines Molecular and Cell Biology* by William D. Stansfield, Jaime S. Colome, and Raul J. Cano, McGraw-Hill, 1996)

DNA SYNTHESIZERS

The technique that you have just learned would enable you to produce and use all bioactive proteins—if living organisms were not a complex system of checks and balances. However, most bioactive molecules are modified after their initial production from the mRNA. The modification requires other proteins that are produced according to other environmental signals and act as a way to fine-tune the action of the hormone.

For example, our old familiar insulin is produced as preproinsulin, stored as proinsulin, and then secreted as insulin. Insulin is produced in is final form by cellular biochemical processess after the gene is transcribed. So, if you produce a protein from the "insulin gene," you will produce preproinsulin. Modification would require the enzymes that are responsible for producing the final form. Methods have been developed to work backward from the amino acid structure of the protein of interest and develop a gene to code for that protein, even though such a gene never existed in nature.

You can synthesize DNA outside of a living cell because the chemical forces that bind the DNA together are operative inside or outside of the cell. To begin the synthesis of a new strand of DNA, you need a matrix to hold the molecules in place, thereby enabling their interactions. DNA synthesizers use a silica matrix in a small column that binds the first nucleotide in the sequence present. The second nucleotide is then flushed through the column and binds to the first nucleotide by means of the sugar-phosphate bond that forms the DNA backbone. The molecules of the second nucleotide have to be chemically blocked so that they will only bind to the first nucleotide and not to one another. The excess of the second nucleotide is flushed out of the column and the chemical blockage removed from the nucleotide that has bound to the matrix. Then the third nucleotide is added and so on.

This process is now performed by a computerized system that, once provided with the desired DNA sequence, will produce the DNA strand. These systems are called *DNA synthesizers*. The variability in the material produced depends on the DNA sequence, but these systems are sufficiently accurate for most proteins of medical interest.

ANTISENSE

Many human disease conditions result because a defective protein is produced. The disease could be treated if you could prevent the gene for this protein from being transcribed, or read out. Also, you might be able to prevent viral infections by preventing the viral gene from being read out. *Antisense* molecules are sequences of organic bases that bind to single-stranded mRNA. The binding of the antisense molecules means that a gene will not be expressed. These strands of organic bases are called antisense, not because they make no sense, but because they counter the "sense" function (the "make sense" portion) of a gene.

An antisense molecule is a single-stranded nucleic acid bearing organic base sequences identical to the coding strand of a gene. As such, this molecule will bind to the mRNA that reads out from the gene because the mRNA is a sequence of organic bases complementary to the gene sequence. The mRNA is only functional as a single-stranded molecule. When the antisense strand binds to the mRNA and creates a double strand, the mRNA cannot produce a protein. In addition to its

potential application in preventing viral diseases and read out of harmful proteins, antisense technology has been used to research the function of various genes. It has potential application clinically as a way to prevent expression of genes that cause pathology, such as cancer-causing oncogenes.

A LIBRARY OF GENES

We are truly, no kidding, living in an information age. In generations past, new scientific developments were shared intermittently at scientific conferences between colleagues or by the occasional publication of the revolutionary scientific paper. Now, the ever-ingenious scientific community has devised ways to communicate small advances, tiny but crucial steps, almost instantaneously using network connections.

A number of scientific institutions are maintaining databases to track and index genetic information as it is derived. These include the Online Mendelian Inheritance I Man (OMIM), a databases maintained by researchers at John Hopkins School of Medicine. This database is devoted to human genes, genetic traits, and disorders. Locus Link is an NCBI databases that serves as an interface to genetic information from a variety of bioinformatic sources. GeneCards was developed at the Weizmann Institute of Science. There are databases for DNA sequences and for protein sequences. There are databases devoted to variability among given alleles. Databases are available to search all the other databases. These and many other tools attempt to consolidate all of the massive information being generated at a daunting rate with the aim of facilitating further progress in the realm of human genetics.

Summary

The most fundamental technology in the biotechnology revolution is the ability to copy a gene from one organism and put it into another. Most remarkably, this can be done between animals of different genus, species, families, and phylums. Where do you get the genes? A methodology exists to go backward from the messenger RNA and rebuild the DNA using an enzyme called reverse transcriptase. The messenger RNA is the molecule dedicated to carrying the code from the DNA to the protein factory. It can be collected from cells that you expect to be expressing the gene you want. Once you have the gene, you can figure out the organic base sequence using a number of analytical techniques. These can involve the use of DNA probes that fluoresce or emit radiation when they attached to their complementary target, or involve digestion with restriction enzymes and the use of gel electrophoresis to separate according to size. Once you know the sequence, you can build the gene using a DNA synthesizer. Once you have a little bit of the DNA you want, you can make more using the polymerase chain reaction (PCR).

To insert the gene into another organism, you need a vector. This vector may be a plasmid, a virus, or an artificial chromosome. An artificial chromosome is built from the basic structure of a centromere at the center and telomeres at either end. The species that provides the centromere and telomere structure gives the species identity of the chromosome. That's how we have bacterial, mouse, and human artificial chromosomes. An artificial chromosome has been likened to a genetic "cassette" that is simply inserted into the recipient and starts acting like a natural chromosome. If you have to insert the gene into an existing vector, you use restriction enzymes. The restriction enzymes make a jagged cut into the DNA of the vector, creating ends that are seeking complementary pairs, called sticky ends. You can stick ends onto your gene that are the complementary pairs that the sticky end of the vector wants. When combined in the presence of ligase, the vector takes up the new DNA. Usually, you will need to include a gene that "reports" the uptake of the gene you want. In bacteria, you can select for organisms that carry a gene for ampicillin resistance and also carry the popular reporter gene, beta-galactosidase. Bacteria containing the beta-galactosidase will cause the media where they grow to change colors. Monoclonal antibodies can be used to confirm that the protein you want is being produced.

Quiz

1. Recombinant DNA:
   (a) is defined as a mix of human and animal DNA.
   (b) contains a segment of DNA from a foreign source.
   (c) is created from embryonic stem cells.
   (d) is required for human cloning.
   (e) is a technique for repairing damaged DNA.
2. Restriction enzymes:
   (a) will dissect DNA at specific nucleotide sequences.
   (b) are naturally occurring.
   (c) exist in great variety.
   (d) cut the DNA so as to create sticky ends.
   (e) all of the above.
   (f) a and d are correct.

3. Sequencing DNA:
   (a) requires specific chemical assays for each nucleotide.
   (b) can be done using a combination of chemicals that stop synthesis at known points and separate segments of DNA of different length by gel electrophoresis.
   (c) can be done using DNA probes that have the "mirror" of the base sequences you are looking for.
   (d) all are correct.
   (e) b and c are correct.
4. Vectors:
   (a) always involve a virus because viruses are able to insert DNA into a cell.
   (b) are a form of genetic material suitable for transport into living cells.
   (c) include artificial chromosomes.
   (d) must include a portion of the host DNA.
   (e) b and c are correct.
   (f) b, c, and d are correct.
5. Artificial chromosomes:
   (a) are constructed using a silica matrix.
   (b) are possible through the use of a DNA synthesizer.
   (c) contain telomeres, a centromere, and a binding site for DNA polymerase.
   (d) are incorporated seamlessly into the host chromosomes.
   (e) all are correct.
6. Altering the genome of a cell:
   (a) may cause chromosomal damage, interrupt normal gene expression, or cause uncontrolled cell growth (cancer).
   (b) requires the use of a vector to introduce new genetic material.
   (c) can be done in the form of artificial chromosomes.
   (d) has been done successfully in human clinical trials.
   (e) b, c, and d are correct.
   (f) all are correct.

7. Microarrays:
   (a) consist of a library of genes to be inserted into host cells.
   (b) are small artificial chromosomes.
   (c) are a collection of DNA fragments from a specific genome, such as a specialized cell.
   (d) are used to determine which genes are being expressed by a specialized cell.
   (e) c and d are correct.
8. Nuclear probes:
   (a) are used to find a given base sequence within a DNA fragment.
   (b) rely on the inclination of single-stranded DNA to mate with single strands with matching (mirrored) base sequences.
   (c) carry either a fluorescent or a radioactive label.
   (d) provide a means to visualize a gene.
   (e) all are correct.
9. DNA ligase is important because:
   (a) it dissects DNA at known base sequences.
   (b) it can be used in conjunction with gel electrophoresis to sequence DNA.
   (c) it induces the union of single-stranded DNA segments with matching base pairs into a double-stranded segment.
   (d) it enables hybridization.
   (e) a and b are correct.
   (f) c and d are correct.
10. The "sticky ends" created by restriction enzymes are important because:
    (a) they allow repair of damaged DNA.
    (b) they enable recombination of DNA and thereby create genetic diversity.
    (c) they provide a means to insert new DNA with matching (or mirrored) base pairs into a given DNA segment.
    (d) they are charged and control the movement in gel electrophoresis.

CHAPTER 8

Proteomics

If you plan on bioengineering cells to produce foreign proteins, there are aspects of protein production and metabolism that you need to know. These include the following:

- In contrast to the adage so familiar to geneticists of a few decades ago, "one gene, one protein," a given gene will produce more than one protein. This explains why the human genome contains only 25,000 to 30,000 genes and can produce 100,000 proteins.
- The functions of proteins are dependent on the shape assumed by these complex molecules.
- The final shape of a protein may depend on the formation of disulfide bonds—bonds between two molecules of sulfur. This type of bond cannot be formed by bacterial cells.
- The shape of a protein depends not only on the protein's chemical structure but also on the action of other molecules present in the cell. These molecules help to fold the protein.

- The shape of a given protein can change. In other words, a protein can be folded and then refolded depending on the needs of the cell.
- The types and amounts of proteins present in a given cell are in tremendous flux, depending on the demands of the environment.

The field of study dedicated to the production and metabolism of proteins is called *proteomics*. The inventory of proteins within a given cell is known as the cellular *phenome*. Many disease conditions are the result of abnormalities in the way that proteins are handled by the cell. The study of proteins and their metabolism after transcription of the gene may yield even more dramatic progress in human health than the study of the genes themselves.

One Gene, More Than One Protein

When the Human Genome Project began, researchers believed that it would entail sequencing of at least 100,000 genes. They thought that each one of the 100,000 proteins produced by human cells would have a correlating gene. When the genome was found to contain only 25,000 to 30,000 genes, it became obvious that each gene could produce more than one protein.

Remember that a given gene sequence on the DNA strand contains some areas that don't code for amino acids. These areas are called *introns*. The noncoding areas are transcribed onto the mRNA with the rest of the sequence and have to be edited out of the mRNA, or else they will disrupt the production of the protein when the mRNA is translated. Some of the enzymes that are active in the editing process are themselves RNA molecules. RNAs with enzymatic activity are called *ribozymes*.

In addition to editing out nonsense areas, cellular enzymes modify the messenger RNA by eliminating specific organic basis in the code and thereby changing the code. The protein produced depends on how the editing is done. A given mRNA molecule can be edited one way to produce one protein and another way to produce another protein. This process of mRNA editing is used to produce different proteins from the same gene.

Structure of Proteins

Before launching into a discussion of proteomics, it is wise to review some of the concepts regarding proteins discussed in Chapter 1. Proteins are composed of *amino acids*. Amino acids are strings of carbon, sometimes formed into a ring, sometimes

Carboxyl group, Amino group, Side chain, (a), (b), (c), $+H^+$

Figure 8-1 Formation of a peptide bond

not. Side chains protrude off of these carbon chains. At least one of the side chains is an amine—NH_3—and another side chain is a carboxyl group—COOH. Amino acids form polymers by linking through a characteristic bond, the *peptide bond.* The peptide bond is formed when the carboxyl group and the amine share a hydrogen (see Figure 8-1).

The key to the proper function of many proteins is the shape. Proteins with exactly the same amino acids, although in different order, can be shaped differently and perform a different function. And, as we will see next, identical proteins may or may not share the same function, depending on how they are folded.

The shape of a protein develops in stages. The structure of the original protein as it is created on the ribosome is known as the *primary structure,* shown in Figure 8-2. The primary structure is a long string of amino acids forming individual polypeptide chains.

Links form between the side chains, causing the protein to folds around itself. The resulting shape is called the *secondary structure* (see Figure 8-3). Additional links support an third layer of folds, called the *tertiary structure* (see Figure 8-4).

The final structure of the protein is the *tertiary structure,* which is formed when tertiary protein chains associate through links in the side chains. The resulting molecule may form a huge ball (a globular shape) or may form pleated sheets. Proteins that act as enzymes tend to be globular (see Figure 8-5). Structure proteins, such as those that form hair or cellulose, assume the pleated sheet shape.

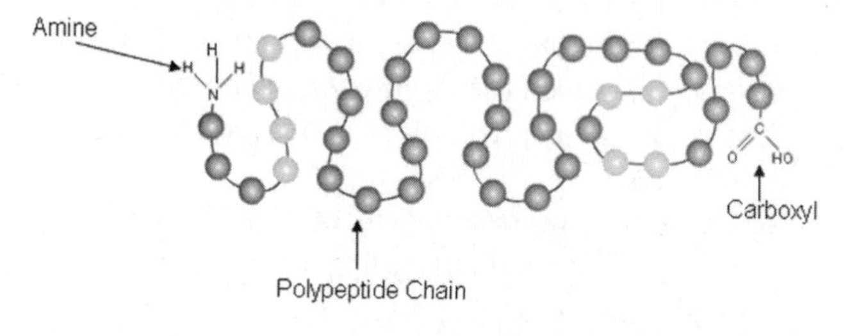

Figure 8-2 Protein primary structure

Figure 8-3 Protein secondary structure

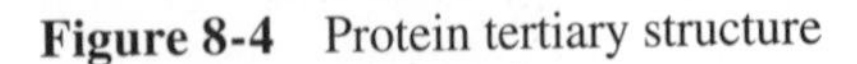

Figure 8-4 Protein tertiary structure

Figure 8-5 Protein quaternary structure

Side chains can link in several ways. They may simply attract one another by their electrical charge or they may form hydrogen bonds, wherein the electronegative pole created by a hydrogen atom in one molecule attracts an electropositive pole on another molecule. A prevalent bond is a *disulfide bond*. Disulfide bond is a covalent bond between two sulfur atoms. Proteins frequently contain sulfide groups as part of their side chains. Remember that a covalent bond is formed when atoms share an electron. Disulfide bonds are formed in the eukaryotic cell cytoplasm. However, conditions within prokaryotic cells ordinarily do not allow these bonds to form. If you are engineering bacteria to produce a protein that requires the development of disulfide bonds, you need to be aware that this protein will not form its quaternary structure. Bioengineers may use mutant bacteria, with different intracellular conditions, or may take the partially formed protein and induce formation of the quaternary bonds outside of the cell.

Protein Folding and Misfolding

The sequence of the amino acids is not the only factor in determining the shape that a protein will assume. There are molecules within the cell environment that actively participate in the folding of a protein.

The collection of molecules that influence the folding of proteins is called *folding moderator,* of which there are several types. Some molecules accelerate the rate of folding and are called *folding catalysts*. Some actually serve to change the shape of the protein and are called *folding chaperones*. There are four types of molecules that act as chaperones: (1) molecules that enable proper folding (*folding chaperones*), (2) molecules that are designed to hold partially folded molecules until the system has the capacity to finish the folding activity (*holding chaperones*), (3) chaperones that refold proteins that have become misshapen (*disaggregating chaperones*), and (4) chaperones that escort proteins to be secreted through the cell membrane (*secretory chaperones*).

Folding chaperones help the protein to fold properly. Many of these are small sugars or short stretches of amino acids. Envision a manufacturing assembly line. As an item moving through the assembly line is produced, you might insert some temporary devices, such as clips or clamps, to hold in a certain configuration through a few steps in the assembly. After these steps are complete, you might remove these holding devices. Farther down the line, additional holding devices might be necessary, only to be removed before the final product is released. The small folding chaperones act just like your clips or clamps in your assembly-line process, holding the device in the proper configuration to complete the next step. If folded improperly, proteins may fail to function or may accumulate into insoluble aggregates known as *inclusion bodies*.

The cell interior is aqueous. Molecules in the cellular fluid usually bear an electric charge—they are hydrophilic or water-loving. Molecules with no electrical charge are hydrophobic, or water-hating. They are called water-hating because they avoid association with water molecules. In the long, linear sequence of a protein, there are areas that are charged, hydrophilic, and areas that have no charge, hydrophobic. In the watery environment of the cell, the hydrophobic surfaces of the protein want to fold inward, leaving the hydrophilic areas to project outward in association with the water molecules of the cell. A typical function of the small folding molecules is to cover the hydrophobic surfaces of the protein, giving these surfaces a charge or covering a charged surface, allowing it to fold inward. By adding and removing these molecules, the cell determines if and when a given hydrophobic surface of the protein folds inward, thereby affecting the shape of the protein (see Figure 8-6).

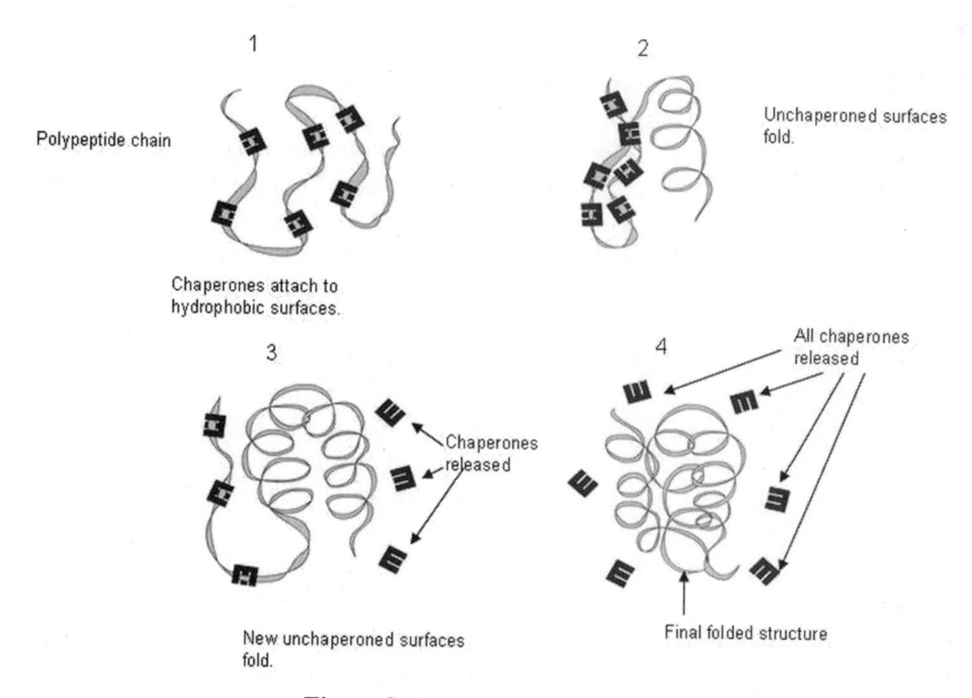

Figure 8-6 Actions of chaperones

Holding chaperones bind partially folded proteins on their surface, serving as a reservoir of these proteins until folding chaperones become available. They will hold these proteins in conditions of chemical or thermal stress, until the environment within the cell is more conducive to the proper folding of the protein. This is one mechanism the cell uses to prevent misfolding and to conserve proteins. The other method is through the action of the disaggregating chaperones. The disaggregating chaperones accomplish the refolding of misfolded proteins. They perform an important quality control function for the cell by salvaging, to the extent possible, proteins that have gone awry. In spite of these salvaging operations, a certain percentage of the cellular proteins ends up in the garbage heap, i.e., in the insoluble inclusion bodies. Inclusion bodies are apparent in the cell as small, dense accumulations, representing clumps of waste proteins.

One feature of chaperones that you as a bioengineer will find particularly useful is that they tend to be relatively nonspecific. In other words, a chaperone molecule will function to aid folding in more than one protein. Researchers investigating conditions caused by misfolding of proteins have randomly introduced into the diseased cells organic molecules that have structures likely to function as chaperones. They have found, by this chance method, molecules that will correct some protein misfolding. Because of the general nature of chaperones, you can introduce various chaperones into your bioengineered system and effect proper protein folding in environments where it would not otherwise occur. The development of specialized chaperones to accurately fold recombinant (bioengineered) proteins is a very active area of bioengineering research.

A protein may be folded more than once. Consider a protein that is destined to reside in the cell membrane. The protein may be manufactured in the cell cytoplasm and need to move to the vicinity of the cell membrane. Then, such proteins will enter the cell membrane, float up through the membrane, and stick their heads out to form a receptor. The protein may require folding in one way by one set of molecules for transport to the cell membrane, only to be refolded prior to actually entering the membrane.

The *periplasm*, an area just beneath the cell membrane, of bacteria contains chaperones that assist the folding and membrane insertion of outer membrane proteins. In eukaryotic cells, most of the post-translational modification of proteins destined for exportation or for insertion into the cell membrane occurs in the lumen of the endoplasmic reticulum or in the Golgi apparatus. These organelles are devoted to the sequestering and manipulation of proteins.

To secrete a protein to the outside of the cell, another control system is required that involves *secretory chaperones*. These secretory chaperones recognize a signal on the protein appropriately called the *secretory sequence*. This sequence is bound by the secretory chaperone, and the chaperone may drive itself into the cell membrane, carrying the export protein along.

Proteolysis—Protein Breakdown

The cellular system that controls protein metabolism includes the capability to destroy proteins and recycle the amino acids. The destruction of proteins is called *proteolysis*. Most proteins that are degraded have undergone multiple cycles of folding and misfolding. Enzymes that dismantle protein polypeptides are *proteolytic enzymes* or *proteosomes*. They are typically barrel-shaped structures that snip the proteins into tiny bits as if the protein had been stuffed into a garbage disposal.

Proteolytic enzymes prevent the accumulation of abnormal polypeptides within the cell and conserve the energy invested in producing amino acids. In eukaryotic systems, the destruction of proteins is preceded by the adherence of markers. These markers are so ubiquitous in eukaryotic cells that they are called *ubiquitin*. Once a ubiquitin molecule binds to the unfortunate protein, many others attach and attract the destroyer proteosome. See Figure 8-7.

Most proteins that are targeted by the proteolytic enzymes have degenerated due to environment stress or were misfolded initially. Some proteins have a built-in obsolescence, i.e., they are engineered to have a short life span.

The flowchart shown in Figure 8-8 and the following steps illustrate the process that the cell uses to produce proteins, to ensure proper folding of proteins, and to destroy proteins.

1. Production of proteins such that proteins are only produced when environmental conditions indicate they are needed.

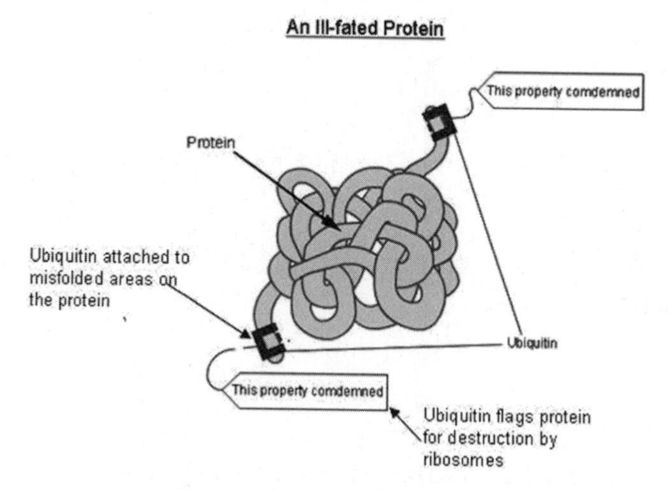

Figure 8-7 An ill-fated protein

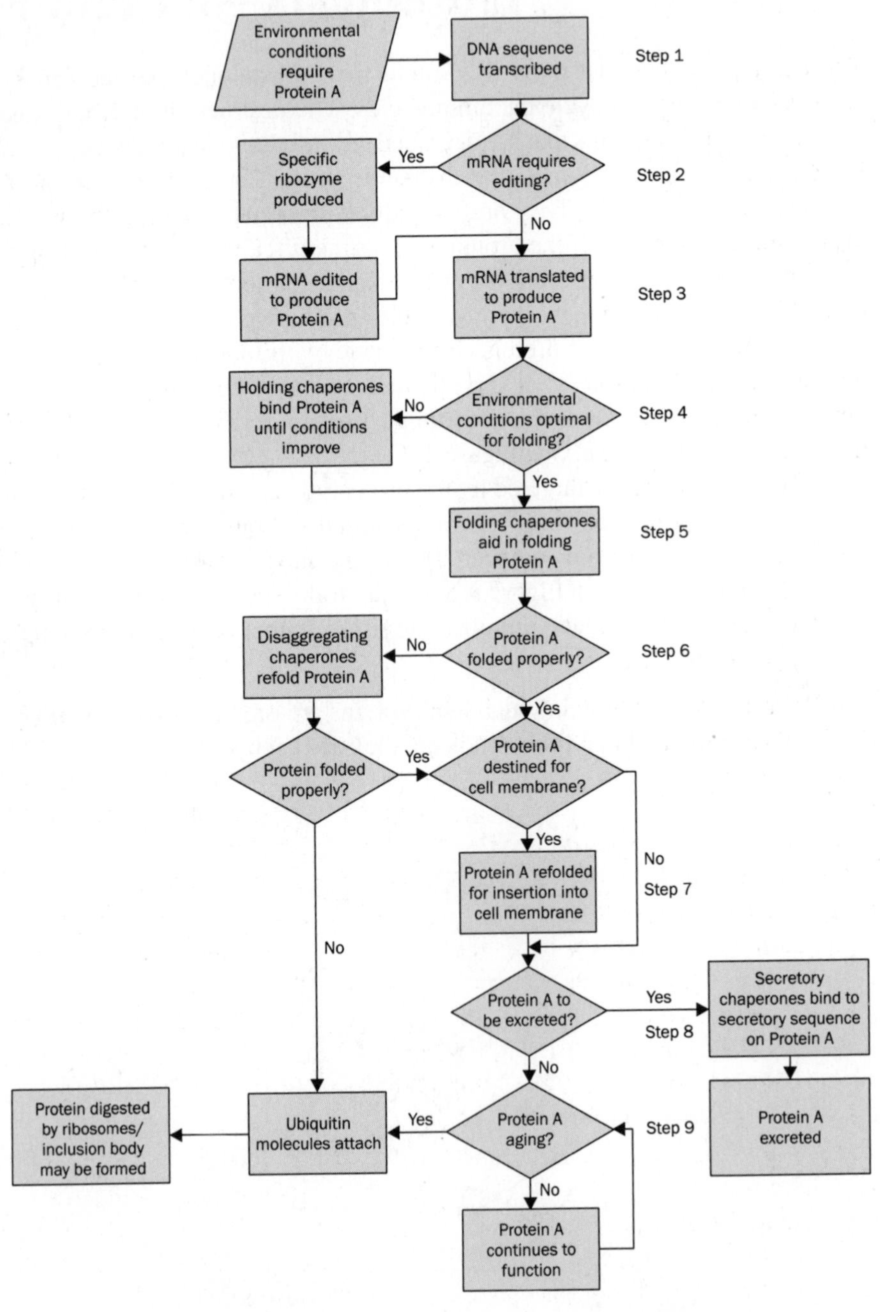

Figure 8-8 Summary of protein production and metabolism

2. Many genes code for more than one protein. The mRNA transcribed from the DNA may contain nonsense code or may need to be edited to produce the correct protein.
3. mRNA is edited by cellular enzymes.
4. After the protein is produced, it may be sequestered by holding chaperones until environmental conditions favor proper folding.
5. Folding chaperones are needed for correct folding.
6. If folding is done improperly, disaggregating chaperones will unfold the protein and allow refolding.
7. Many proteins must be refolded to enter the cell membrane.
8. Sequestering chaperones must bind to special sequences on proteins before they can be excreted outside of the cell.
9. Aging proteins or proteins that are hopelessly misfolded are flagged by the attachment of ubiquitin and are digested by ribosomes.

Reducing Protein Misfolding in Bioengineered Systems

This issue of protein folding and misfolding is going to complicate your life, bioengineer, because you cannot necessarily insert your gene for your protein into bacteria, or whatever, make provisions to ensure that it is translated and transcribed, and go on your merry way. You need to concern yourself with the way the protein is folded. There are steps you can take to reduce misfolding.

You, bioengineer, are inducing the bacterial cell to do a very unnatural thing—produce eukaryotic proteins. To do so, as you recall, you insert the DNA sequence into the bacteria using a vector, such as a plasmid or a virus, and you allow the cell mechanisms to transcribe the DNA sequence for your protein. This protein is foreign to the cellular environment, and vice versa. Not unexpectedly, protein misfolding is a significant problem. There are several strategies you may employ to control this problem. The first is to reduce the synthesis rate. Protein overproduction is a significant stress to the cell and will increase misfolding. You can slow the cell down by changing environmental conditions, such as reducing the temperature of the culture. Another strategy is to also insert the genes to make folding modulators. These genes for the folding modulators will be expressed along with the target protein genes. If the presence of folding modulators in the cell is increased, the efficiency and accuracy of protein folding protein may also be increased.

You can also increase the efficiency of secretion of your molecule by inserting the genes for the chaperones involved in secretion. Again, these chaperones help to move the protein outside of the cell. Once outside the cell, you, bioengineer, can collect this protein for your own use. Recall that proteins to be excreted have special sequences of amino acids on them. These sequences are bound by the secretory chaperones. The secretory chaperones then tug the protein through the cell membrane. By ensuring that the signal sequences are part of your inserted gene, you ensure that the protein can be excreted to the outside of the cell.

Protein misfolding increases with certain environmental factors, such as heat stress. Also, as the rate of protein production increases, so does the rate of misfolding. As a bioengineer of proteins, you strive to set up production systems to maximize production. You have to balance your desire for a high production rate with the loss you will sustain with increased misfolding of your proteins.

Famous Misfolded Proteins

New research into how proteins are folded, misfolded, and refolded has promise in the treatment of human conditions associated with the accumulation of misfolded proteins. One of these conditions is aging. As we get older, our cells and extracellular spaces tend to accumulate insoluble proteins. These proteins have lost their globular shape and appear as insoluble, sticky fibers called *amyloids*. The excessive accumulation of amyloids can cause diseases such as early-onset Alzheimer's, Parkinson's, early-onset cataracts, alpha-1-antitrypsin deficiency, Type II diabetes mellitus, and systemic amyloidosis.

Several diseases that are popular with the media are due to the accumulation of *prions*. These diseases are in a class of transmissible spongiform encephalopathies (TSEs), a name earned by the fact that they turn brain tissue into sponge. These include bovine spongiform encephalopathy (mad cow disease), scrapie (affects sheep), Creutzfeldt-Jakob disease (CJD), and Kuru (risk factor for individuals who consume human brain tissue). Prions are aggregated proteins that trigger aggregation of further prions. The prion protein switches from its soluble alpha helix form to a partly beta sheet structure when it encounters another aggregated prion. The aggregations form long fibers in the cell, causing cell death. To give you an idea of how truly hardy prions are, you can boil them or digest them in acid, with little effect.

Research into Protein Type and Function

The *phenome* is the repertoire of all of the proteins in a cell. Attempts to describe the phenome of given cell types occupy a large amount of research. Even with

modern computerized technology, characterizing the phenome is a laborious process and is unlikely to flag rare proteins. A goal of proteomics is to produce microarrays similar to what is available as cDNA chips. Remember that the DNA chip contains the genes of a given cell on a well-mapped matrix. The genes bound to the matrix are denatured such that only one strand of the double-stranded DNA is on the chip. The single strand will bind a complementary strand. You can link a label to the complementary strand and see the genes on the chip. However, there is no such thing as a complementary strand to a protein. The identification of the various proteins on the chip would require specific monoclonal antibodies against each and every protein that might be on the chip.

Determining the functions of the incredible myriad of proteins in the phenome is a big job, one that will occupy bioengineers for the foreseeable future. Historically, this has been done by comparing phenomes of mutant and normal cells and individuals, or analyzing biochemistry of individual proteins, and comparing the structure of proteins under investigation to that of proteins with known function. The emerging science of sequencing DNA has been helpful because genes for proteins of similar function tend to be located close to one another. In some cases, the genes are located together in more primitive animals and have drifted apart in more highly evolved species. The ability to turn off genes using antisense technology will reveal function by illustrating the effect of depriving the cell of said function. Antisense technology deactivates mRNA by providing strands of RNA complementary to the mRNA. The complementary strands bind to the single-stranded mRNA, preventing it from producing a protein. Large databases have been developed to aid in this type of analysis. Among these is a network of protein-protein interactions in yeast, developed by Schwikowski et al. (2000). Analysis of such networks can uncover function of uncharacterized proteins.

Summary

The inventory of cellular proteins is known as the phenome. Study of the phenome is the science of proteomics. This is an extremely complex field due to the huge variety of proteins and the changing landscape with changing environmental conditions. Proper function of proteins depends not only on the correct transcription of the code but also on the proper execution of the folding of the protein. Protein folding has been determined to be an active process and under the control of regulatory molecules known as folding modulators. The folding modulators have a relatively nonspecific function, which is to cover up specific surfaces until a certain point in the folding sequence. Recent research on the mechanism for the proper folding of proteins has revealed that some human disease conditions are not due strictly to defective gene product but due to improper folding of this product after production. Also, the function of secretory modulators has been studied. Improved protein secretion could be important commercially for production systems where secretion of the protein is a limiting factor.

Quiz

1. Protein folding:
   (a) is completed when the polypeptide chain folds around on itself.
   (b) is governed solely by the chemistry of the protein.
   (c) is completed by the time the protein is released from the ribosome.
   (d) all of the above.
   (e) none of the above.
2. Folding modulators:
   (a) tend to bind to hydrophobic sections of the protein.
   (b) tend to be highly specific to distinct proteins.
   (c) are hardy, stress-resistant molecules.
   (d) are found only in eukaryotic cells.
   (e) b and c are correct.
3. Holding modulators:
   (a) bind misfolded proteins for destruction.
   (b) will hold partially folded proteins in times of environmental stress.
   (c) release proteins when folding modulators become available.
   (d) all are correct.
   (e) b and c are correct.
4. If a protein is marked with ubiquitin,
   (a) it will undergo proper folding.
   (b) it will be held until folding moderators are available.
   (c) it will be protected from environment stress.
   (d) it is destined for destruction by proteases.

5. Misfolded proteins:
   (a) are often refolded to the correct configuration.
   (b) can cause disease, either by failure to function properly or by causing excessive aggregates to form in the cell.
   (c) are recognizable by the presence of hydrophobic structures on their surface.
   (d) typically cannot be picked up by secretory chaperones and cannot be extruded or inserted into the cell membrane.
   (e) all of the above.
6. Amyloids:
   (a) are rare in individuals without debilitating diseases.
   (b) accumulate in the neurological tissues of individuals with early-onset Alzheimer's disease.
   (c) are a consequence of protein misfolding.
   (d) can cause cell death.
   (e) all are correct.
   (f) b, c, and d are correct.
7. Cellular phenomes:
   (a) can be displayed on a microchip.
   (b) consist of proteins, lipids, and carbohydrates.
   (c) are encoded on the cellular DNA.
   (d) can be inserted into other cells using a vector.
   (e) all are correct.
8. Formation of disulfide bonds:
   (a) is not supported in natural prokaryotic cells.
   (b) is necessary for the production of many commercial attractive proteins.
   (c) can be done outside the cell.
   (d) all are correct.
   (e) b and c are correct.

9. Proteins:
   - (a) are folded by ribosomes.
   - (b) assume a secondary shape that includes complex folding regimens.
   - (c) can preserve their function even if only partially folded.
   - (d) can form insoluble aggregates inside the cell known as inclusion bodies.
   - (e) all are correct.
10. Folding moderators:
    - (a) sometimes fold proteins more than once.
    - (b) sequester proteins during times of environmental stress.
    - (c) include molecules that help in secretion of proteins.
    - (d) all are correct.

Reference

Schwikowski, B., P. Uetz, and S. Fields. 2000. A network of protein-protein interactions in yeast. *Nature Biotechnology* 18:1257–1261.

CHAPTER 9

Stem Cells

The stories coming out of the stem cell research world are fantastic. Reports include animals with spinal chord injuries made to walk again, diseased organs healed, diabetes cured, and animal models of Parkinson's disease successfully treated. Stem cells really do hold these promises, and more—they stand to revolutionize medicine and may change the way we think of life and death altogether. Certainly, stem cell therapy may change the way we view the limitations that nature has imposed upon our bodies.

Stem cell research has attracted the attention of the public both because of the potential for human health and because of the ethical implications. In this chapter, we will review the status of stem cell research as it exists today so we will be better able to address the changes that it may bring tomorrow.

Somatic Cells, Germ Cells, Stem Cells, Adult Stem Cells, and Embryonic Stem Cells

Before we launch into a discussion of stem cells, we need to review some basic categories of cells in a multicellular organism. In classic biology, cells are considered either *somatic* or *germ* cells. Somatic cells compose the tissues of our bodies, contain two copies of each chromosome, and are known as *diploid* (two copies of everything). Germ cells are either egg or sperm. They contain only one copy of each chromosome and are known as *haploid* (one copy of everything). The union of germ cells in the fertilized egg restores the genetic material to diploid conditions—two copies of everything. Somatic cells are typically highly *differentiated*, or grown up. Differentiated cells are usually highly specialized cells and are fully developed. They are committed to their role in life, and their appearance and cellular chemistry are devoted to their particular function. For example, muscle cells have a distinctive appearance, contain special fibers to allow contraction, and couldn't be confused with any other type of cell. *Undifferentiated* cells are more primitive and do not perform high-level functions. Their appearance is generic and they have no specialized functions.

Relatively recently, researches have discovered that cells that have not differentiated (grown up) are living among the differentiated somatic cells in the organs of the body. These undifferentiated cells retain the capability to specialize, to chose an occupation if you will. They are called *stem cells*. By definition, stem cells have the ability to reproduce indefinitely. They also retain their ability to differentiate into specialized cells. However, the stem cells found in a given organ of the body appear only to be able to differentiate into cells found in that organ.

The first such cells to be recognized and used therapeutically were found in bone marrow. These stem cells can become monocytes, lymphocytes, neutrophils, basophils, and erythrocytes. They generate blood cells. Cells that give rise to blood cells are known as *hematopoietic* cells, so stem cells found in bone marrow are called *hematopoietic stem cells* (HSCs). Patients receiving a bone marrow transplant benefit from the stem cells that are in the transplant because these cells diversify to fill the needed population of the various types of blood cells.

Stem cells have also been isolated from fetal tissue and from umbilical cord blood. Stem cells existing in tissues are considered *adult stem cells*, simply because they are not derived from developing embryos; stem cells derived from embryos are called *embryonic stem cells* (ESCs). Fetal and umbilical cords stem cells have been used experimentally in the treatment of several disorders. They appear to be more versatile than other adult stem cells and less likely to induce an immune response.

Embryonic stem cells come from fertilized eggs. Like the adult stem cells, the embryonic stem cells are diploid. They have a full complement of DNA. When an egg is fertilized, it divides to form a mass of cells called a zygote. Embryonic stem cells are

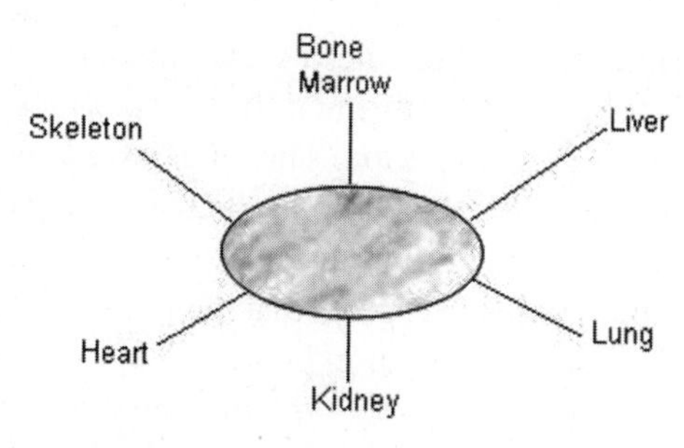

Figure 9-1 Embryonic stem cells

obtained early enough in the life of the zygote that they retain the capability to become any of the types of cells needed by the complete organism. This must occur very soon after fertilization. By the time the zygote travels down the mother's fallopian tube and implants into the uterus, the cells of the zygote, although still just a ball, have generally committed to their lot in life and are no longer embryonic stem cells. (See Figure 9-1.)

The ultimate stem cell is the fertilized egg—it can develop into the entire complex organism and hence is called totipotent. Embryonic stem cells obtained early in the development of life can give rise to any type of cell in the body, but an embryonic stem cell cannot to be induced to form an intact organism, at least not under current technology. They are, therefore, called multipotent. Adult human beings also have stem cells; however, these cells are more specialized than the embryonic stem cells in that they are devoted to become cells of a particular organ. For example, the HSCs can become any of the cells of the bone marrow, but their potential to become liver or kidney has yet to be demonstrated. These cells are pluripotent. Pluripotent stem cells have been isolated from the brain, muscle, skin, GI tract, cornea, retina, liver, and pancreas. They apparently exist to provide organs with some ability for self-repair.

Adult Stem Cells

Adult stem cells have been derived from adult tissues, from umbilical cords, and even from fetal tissues. Note that all of these are considered adult stem cells even though their sources are not necessarily adult tissues. Stem cells, like all cells in the body, contain the genetic instructions to do all of the cellular functions. Theoretically, any undifferentiated cell could become any of the 200 cell types in the body. What causes cells to go in different directions as they mature is not completely understood.

Because adult stem cells have not yet begun to differentiate, the hope is that they can be convinced, by whatever mechanisms control the direction of maturation of a cell, to become many types of mature cells. We would like to take bone marrow cells and induce them to make heart cells, for example. There are laboratory reports that adult stem cells have the capability to become a wider range of cells than previously thought. HSCs have been injected into mice that have widespread tissue damage due to exposure to high levels of radiation. The HCS have been reported to migrate into tissues other than bone marrow and repair damaged lungs, livers, and kidneys. However, this type of research is preliminary and mice cells may be very different than humans. For example, there are reports that the number of active genes in stem cells from mice is significantly less than in humans.

Adult stem cells are currently the only type of stem cells routinely used to treat human diseases. However, their uses to date are fairly limited. Embryonic stem cells have not been used clinically. Stem cells in bone marrow transplants are used to treat victims of blood cancers and also of hemophilia. Stem cells are important in the growth of skin grafts for plastic surgery for severe burn victims. A portion of the victim's skin is removed and maintained in culture where skin stem cells produce additional skin and enlarge the original tissue. The resulting graft is called *autologous*, because it comes from the victim. Clearly, the surrounding graft provides the correct signals to the stem cells to grow and differentiate into additional skin.

Bone marrow stem cells have been injected in patients with diseased hearts in clinical trials. Clinicians report modest repair of the damaged heart. However, there is no evidence that the stem cells participated in the repair. The stem cells may secrete growth promoting factors and contribute to the formation of new blood vessels. Stem cells from fatty tissue and bone are used to repair cartilage injury in horses. Stem cells derived from fat tissue are under investigation for breast reconstructive therapy.

Stem cells from umbilical cord blood have shown promise in clinical trials as sources of HSCs. These cells appear less likely to invoke an immune response from the patient than are HSCs from adult sources. Also, cells from fetal tissues have been used in clinical trials to treat Parkinson's disease and diabetes. These trials have reported mixed successes. Certainly, clinical successes with adult stem cells will increase as the understanding of the phenomenon of cell differentiation increases.

The hope is that adult stem cells can be used to repair damaged organs. Theoretically, stem cells could be induced to develop whole new organs that could then be used to replace damaged organs. There are several technical obstacles to this therapy. For one, these stem cells are very rare and difficult to isolate. You might envision finding them, isolating them, and then allowing them to divide in the laboratory. A very daunting obstacle is the fact that they are extremely difficult to grow in culture. It is currently not possible, therefore, to remove them and induce them to increase their population.

Adult stem cells, shown in Figure 9-2, present other difficulties for therapeutic uses. If, for example, liver stem cells are removed from the victim of liver disease and then reinjected to treat the disease, the stem cells may exhibit the same genetic defect that caused the disease in the first place. If stem cells from other adults are used, they may be rejected by the recipient's immune system. Also, because these cells have been present—possibly as long as the individual has been alive—cells

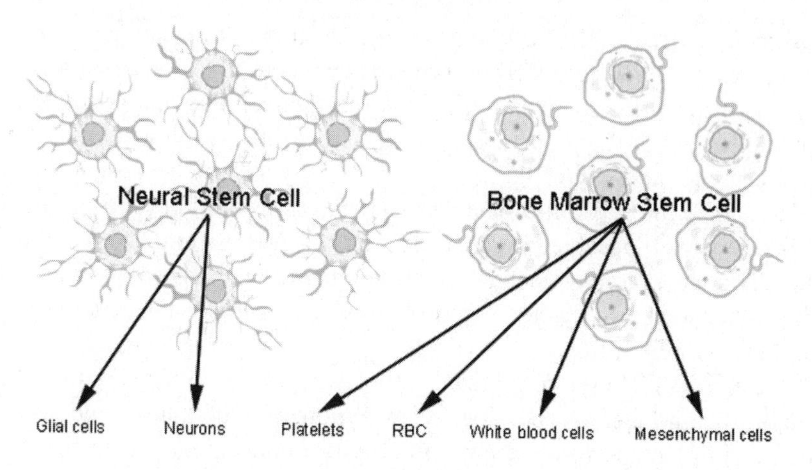

Figure 9-2 Adult stem cells

from older individuals are at risk for DNA abnormalities, due to accumulations of environmental insults such as toxins. Also, unfortunately, researchers report that stem cells decline in viability as either the donor or the patient ages.

Embryonic Stem Cells

Embryonic stem cells offer a huge advantage over adult stem cells in that they can be grown in culture. Embryonic stem cells are simply cells of an early stage embryo. They are removed from the embryo when it exists as a ball of approximately 150 cells, called a *blastocyte*. Only 30 of the cells dwelling in the interior of the ball are destined to become part of the fetus and can be used to establish an embryonic cell line. In normal embryonic development, stem cells disappear after the seventh day of gestation, that is, they can no longer form all of the various types of tissues.

Any one of the embryonic stem cells can be theoretically coaxed into developing into any of the approximately 200 specialized cells of a multicellular organism. Theoretically, embryonic stem cells also could develop into the entire organism, although this has never been demonstrated. In the uterus, however, multiple births of identical individuals are a result of the first few cells of the zygote breaking up early in the pregnancy and each cell becoming an intact individual.

DEVELOPING EMBRYONIC STEM CELLS LINES

Embryonic cells are harvested from donated eggs, usually those that were fertilized as part of an *in vitro fertilization* (IVF) process. IVF is used to enhance fertility by removing sperm from the male and egg from the female, allowing fertilization to occur outside of the body, and then implanting some of the resulting zygotes into a receptive uterus, usually that of the egg donor. We say *usually* because it certainly does not have to be the egg donor. The first IVF baby was born in 1978 in England, and IVF is now a well-accepted practice. The use of IVF has allowed many previously infertile couples to have babies. In addition, IVF is increasingly being used for reasons other than remediation of problems with fertility, such as to allow sex selection, for example.

Commercially, IVF is used to produce cattle. A zygote is obtained in the laboratory through IVF and then frozen. The frozen zygotes can be transported long distances before implantation into a cow much more conveniently than the calves could be moved after birth by their natural mother.

Many eggs must be fertilized to produce a single IVF baby. Typically, up to eight eggs are fertilized per session, and only two or so of these will be implanted into the prospective mother. The rest are frozen in liquid nitrogen when they have reached the six-to-eight cell stage. There are an estimated 400,000 frozen embryos in the United States. Of these, 2.8 percent have been donated to research. The effect of the freezing process on the embryo is unknown, and many embryos do not survive the freezing. However, many frozen and subsequently thawed embryos have developed into viable babies.

To establish a stem cell line, the researcher induces the thawed zygote to grow to the blastocyte stage before teasing out the stem cells. Obviously, the embryo cannot survive this process. Embryonic stem cells are "immortal," that is, they will grow and divide indefinitely. A cell line is kind of like the sourdough in your refrigerator. If you tend it carefully, you can take bits of it to start new batches of sourdough. These in turn can be used to start other batches of sourdough. Some human ESCs have undergone 300–400 duplications. Mouse ESCs have existed in culture for several decades.

One of the features of cell lines that researchers are concerned about is the fact that cell lines tend to "play out" and must be replaced with new cells. This is expected based on observations with other types of cell lines that have been maintained in culture for long periods of time, such as human fibroblasts. Probably, new embryonic cell lines will need to be continuously developed to replace older ones that are no longer dividing. Also, in other types of cultures, cells that have been maintained in culture for a long time accumulate mutations that decrease their viability.

Human embryonic stem cell (HESC) lines are very fragile, so establishing them is quite a feat. (See Figure 9-3.) They tend to want to develop into a more differentiated

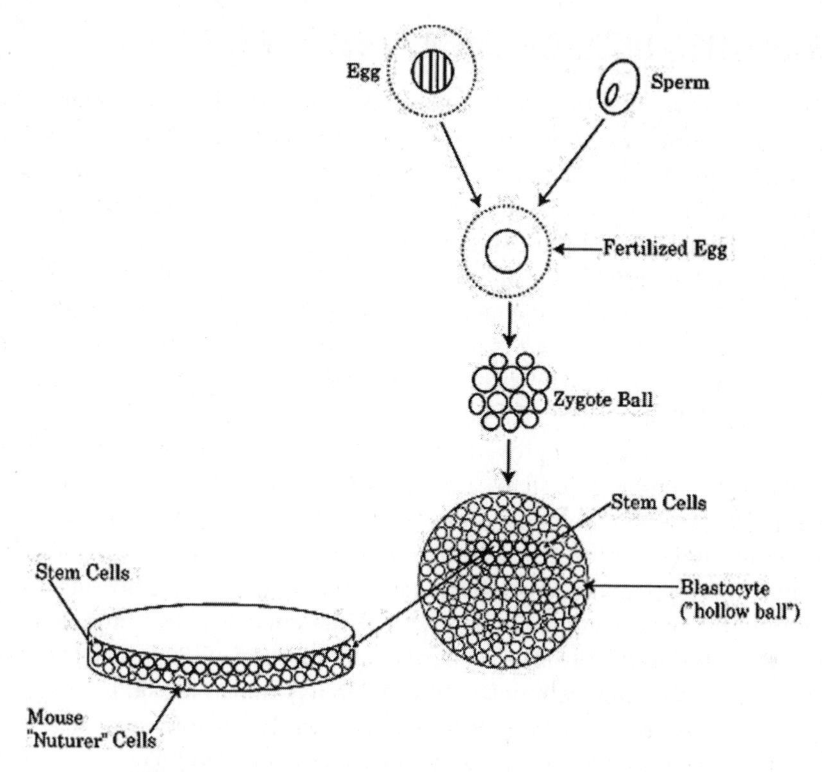

Figure 9-3 Establishing human embryonic cell lines

state. In current technology, they are prevented from differentiating by maintenance on a substrate formed of mouse embryonic fibroblasts and by growing in the presence of cow serum. The need for such "feeder cells" is a bit of problem when you are intent on injecting HESCs into people for therapeutic purposes. This is because the human cells will be contaminated by products or viruses from the mouse cells and/or cow serum. This year researchers have announced an ability to replace the animal nutrients for growth of ESCs with similar human components. However, potential contamination of many existing HESC lines remains.

If removed from the feeder cells, HESCs ball up into what are called *embryonic bodies* and proceed to differentiate rather randomly—producing bone, muscle, and heart tissue cells—all in a ball. If injected into mice at this stage, they form a type of tumor called a *teratoma*. Teratomas are tumors that sometimes develop in the uterus after the death of a fetus. They consist of a random mix of tissues of various types. The fact that the HESC ball does not differentiate in an orderly fashion to produce an organism is evidence that somehow the orchestration of cell differentiation has degenerated in the embryonic stem cell culture.

THERAPEUTIC USES OF EMBRYONIC STEM CELLS

There is no question that the use of embryonic cells promises to revolutionize the treatment of some human diseases. Imagine the marvel of observing a collection of these cells develop into heart muscle cells, coagulate, and begin to pulse, like normal heart tissue. As you can imagine, these cells might be used to repair a damaged heart. However, research using embryonic stem cells is very, very new. We must appreciate the very long lead time between a new therapeutic concept, such as the use of HESCs, and the implementation of this concept into our medical practices. The current state of the art is relatively primitive, and we have a lot to learn.

The research into any human disease typically begins with the development of an animal model of that disease. We have models of heart disease in mice and pigs. The diseased animals have improved when their hearts are injected with embryonic stem cells. Mouse embryonic stem cells have been studied for decades, and methods to force their differentiation into various types of tissues are fairly well understood. Mouse studies have confirmed that ESCs can alleviate the symptoms of diabetes, Parkinson's disease, and spinal cord injury. Recently, the bioengineering company that partially funded the development of the first stem cell line, Geron, announced plans to apply for clinical trials for neurological diseases. Geron owns the patents for 9 of the 22 embryonic cell lines used a federally-funded research.

There are, however, some serious problems with therapeutic uses of ESCs. For example, there remains a potential that the injected cells will specialize in an inappropriate direction, forming kidney or bone tissue instead of heart tissue, for example. There is also the concern that the cells will serve as the seed for cancer. Stem cells are postulated to be the source of at least some cancers *in vivo*. To ensure differentiation in the appropriate direction and to decrease the risk of cancer, projected therapeutic uses of HESCs may need to employ cells that have been induced to partially differentiate.

Another concern is that if the ESCs are antigenically very different from the patient, the patient's immune system will seek and destroy them—just like any other foreign cells. However, ESCs are reputedly less antigenic than their adult counterparts, so this might not be a major complication. In addition, researchers have suggested that the HESCs could be genetically engineered so that they do not express cell surface antigens, thus reducing their antigenicity. However, this technology is still in the earliest stages of research, and it will be quite some time before such cells can be produced.

Somatic Cell Nuclear Transfer—Cloning

Somatic cell nuclear transfer (SCNT) is a method of developing a cell that behaves like an embryonic stem cell but contains the genetic material derived from an adult cell. The process begins with a fertilized egg and a somatic cell from an adult. The

donor of the fertilized egg and the donor of the somatic cell do not have to be the same individual. The nucleus is removed from the fertilized egg and is replaced by the nucleus from the somatic cell. This somatic cell can be theoretically any type of cell, from bone marrow to skin. The egg is induced to develop into a zygote and an embryonic cell line is derived from the blastocyte, just like with any embryonic stem cell line development. However, theoretically, you could implant the new egg, with the somatic cell nucleus, into a uterus and it will develop into an organism—a clone of the individual who donated the somatic cell.

SCNT has been used extensively to clone animals. In 1996, Dolly the sheep was born as the first successful clone derived from adult cells. Since then, scientists have cloned thousands of cattle, mice, and other animals using similar technology. The cloning of Dolly touched off a debate about the meaning of life that still rages today.

In February of 2004, a group led by veterinarian Woo Suk Hwang and gynecologist Shin Yong Moon of Seoul National University shocked the scientific world by reporting the first derivation of ES cells from human nuclear transfer experiments (*Science,* 12 March 2004, p. 1669). This group reported that it was able to remove the nucleus from an undifferentiated human oocyte (egg) and replace it with the nucleus from a differentiated adult cell. The altered cell was subsequently said to be used to establish an embryonic cell line. According to the researchers, these efforts yielded just one cell line from more than 200 tries. However, in May of 2005, this same team reported creating nearly a dozen new lines of human embryonic stem cells, some of which were reportedly derived from diseased or injured patients. They had many, many people volunteer to become the source of a new embryonic stem cell line. Then in the fall of 2005, this entire effort was revealed as a fraud. Woo Suk Hwang resigned in disgrace, and we all went away with a renewed appreciation for the technical challenges remaining before somatic nuclear cell transfer becomes a reality for human cells.

On the American continent, Harvard scientist reported in August of 2005 that they also had created a stable cell line by fusing somatic and embryonic cells. However, their cell line contains the DNA from both the somatic and the embryonic cell lines. Nonetheless, they demonstrated that the cell line would differentiate into the three basic types of human tissues from which all tissues in our bodies are derived.

This development certainly show that somatic cell nuclear transfer technology, shown in Figure 9-4, may someday be available for the development of human cell lines that can be used to treat disease. How does this technique differ from cloning? A clone would be developed only if the modified germ cell (egg) were to be implanted in the uterus and allowed to develop into a baby that is genetically identical to the individual who donated the somatic cell. No serious researchers in the field have an interest in using this technology to clone individual human beings.

The therapeutic potential for embryonic stem cell lines derived in this fashion is enormous. An embryonic cell line could be developed using somatic cells from an individual possessing a diseased organ. Theoretically, the embryonic cells could

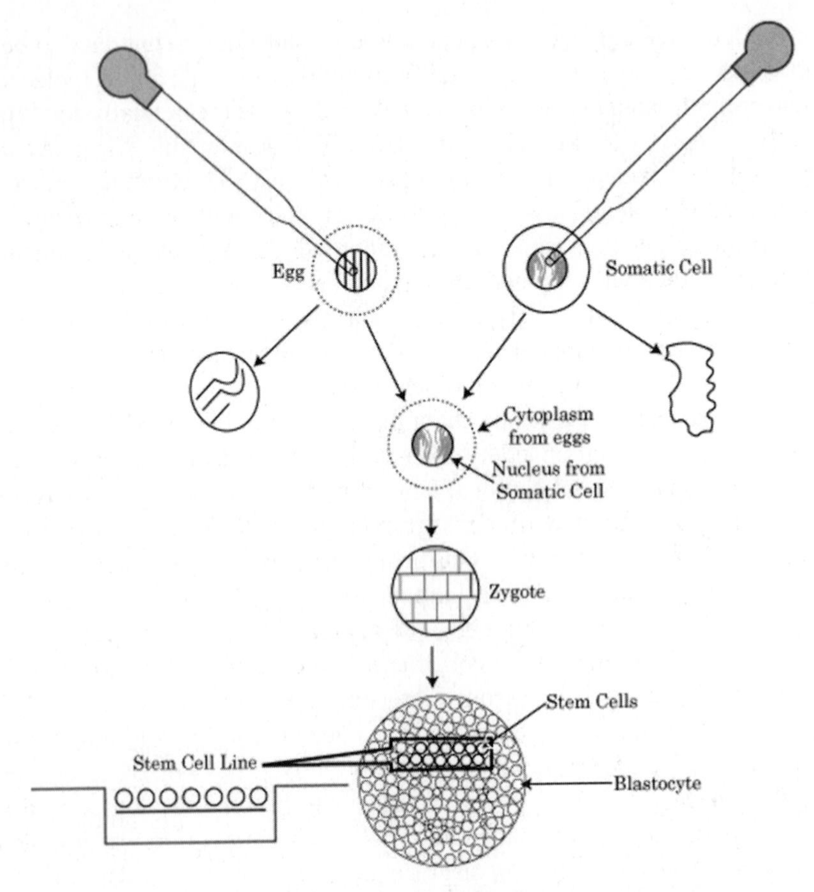

Figure 9-4 Somatic cell nuclear transfer

subsequently be injected into the diseased organ and aid in its repair. More futuristically, the embryonic cells could be induced to form a completely new organ and this organ transplanted into the patient. Because these cells would be genetically identical to the patient, there would be no concern about an immune response to the new organ. Obviously, an increased knowledge of the controls regulating cell differentiation will be needed before the formation of entire organs using SCNT is feasible. Also, if the individual is suffering from a genetically-linked disease, the embryonic cells containing his or her genome would contain the same defect.

The use of the SCNT technology carries some caveats. Animals cloned using SCNT experience a high rate of late-term miscarriages and severe birth defects, indicating that the cell transformation is not without flaws. Also, our society has not confronted the ethical issues with a technology that theoretically could be used to clone human beings.

Controversy and Legal Constraints

In 1996—well before the first embryonic cell line was established—the U.S. Congress banned the use of federal funding for embryo-destroying research. The first embryonic stem cell line in the United States was therefore established with private funding. This was done relatively recently, in 1998, by a group led by Dr. James Thomson at the University of Wisconsin. The research was funded by the Wisconsin Alumni Research Foundation (WARF) and Geron, the company that now owns the commercial rights on numerous HESC lines.

Federal funds to support HESC research were established under the George Bush administration and have only been available since August 9, 2001. However, in accordance with the policy established by President George Bush, this funding is available only for cells established before the August 9, 2001 date. There are 22 human embryonic stem cell lines that meet this federal criterion. All of these stem lines are owned by individual providers, who hold patents on these lines. In spite of this restrictive policy—or perhaps because of it—private interests continue to support HESC research efforts in the United States, allowing development and use of other stem cell lines. Additionally, some states have also provided government funding to offset the lack of federal funding. U.S. policy on the scientific use of HESC varies dramatically from state to state: California for example, provided $3 billion in state funding for stem cell research in 2005; in contrast, Arizona and Pennsylvania have decreed that the creation of a human embryonic stem cell line is a felony.

U.S. interests still own more stem cell lines than any other country. Of the approximately 130 HESCs worldwide, 70 are owned by U.S. companies or universities. However, numerous foreign governments, including Great Britain and South Korea, are actively funding HESC research. Thus, current U.S. government restrictions make it unlikely that the United States will continue to lead in this area of research.

Researchers are interested in the use of federal funding to develop new cell lines for several reasons. First—as described earlier—these cell lines have a tendency to play out, meaning that soon the currently approved cell lines will likely be rendered useless.

Second, the 22 approved cell lines contain the genomes of 22 individuals, which is a woefully inadequate representation of the human genetic diversity in the United States, let alone worldwide. Researchers would like to explore the differences between HESCs from a wide variety of different populations to better understand the genetic predisposition to various diseases, including heart disease, cancer, and many other significant diseases.

Third, the need to differentiate between federally funded research and privately funded research has created a bureaucratic nightmare for researchers. Researchers have to separate the facilities used for the research supported by different sources of funding and are burdened with extensive record-keeping responsibilities.

Fourth, private sector involvement, with its usual concerns about proprietary information, has impeded the flow of information between researchers in this new and burgeoning field and, thus, has undoubtedly had an adverse affect on the progress of research. Some of the patents are unbelievably broad. For example, the Wisconsin Alumni Research Foundation has a patent for "a method of culturing human embryonic stem cells and composition of matter which covers any cells with the characteristics of stem cells." So far, this broad patent has not been used to inhibit the research efforts of others, but the potential exists that future research could be restricted, due to commercial concerns and secrecy. The very fact that private companies hold patents for the cell lines has the potential to severely constrain access to information gleaned from research results obtained using these cell lines. The fact that some of these patents are held by relatively small companies without large R&D budgets has slowed development on some fronts. Thus, the primary use of private funds to carry out stem cell research is not practical in a scientific community where the free exchange of ideas and results is a key component driving the process forward.

Finally, bringing HESC research under a federal umbrella will allow development of an ethical framework for the work, and provide federal oversight of the development and use of HESCs. A model might be the National Institute of Health's Recombinant DNA Advisory Committee (RAC). The RAC was formed because of the concerns of both the public and the scientific community when recombinant DNA technology was first developed that this technology might be misused. The RAC is a panel of up to 21 national experts in various fields that advises the NIH Director and the NIH Office of Biotechnology Activities (OBA). The RAC reviews all research proposals involving human gene transfer. The federal limitations on HESC research are being revisited as of this writing.

Summary

Stem cells can serve as the source of other more highly differentiated cells. Adult stem cells have been isolated from a number of body tissues and exist apparently to provide a repair mechanism for a specific organ. Adult stem cells have been used clinically in bone marrow transplants and to aid in forming skin grafts. Clinical trials with other forms of adult stem cells have had moderate success.

Human embryonic stem cells (HESC) may differentiate into a large number of cells crossing tissue types. These are derived from embryos formed during the IVF process. Their derivation results in the destruction of the embryo. HESCs have enormous therapeutic potential as demonstrated in animals. Researchers report that animal models of spinal cord injury, Parkinson's disease, and diabetes have responded to embryonic stem cell therapy.

Somatic cell nuclear transfer is the process by which the nucleus of an embryonic cell is replaced by the nucleus of a somatic cell. The embryonic cell then becomes a

clone of the somatic cell and forms embryonic stem cells. This is the process used to clone animals and has the potential to produce embryonic stem cells that are genetically identical to the individual who might benefit from therapeutic uses of these cells.

The rapid advances in the area of stem cell research have lead to wild speculation as to the medical future of mankind. However, this science is not as far along as you may have been led to believe. Its main potential is in the revelation of basic cell phenomena such as the regulation of differentiation. From this basic knowledge, as it accumulates, will come the more revolutionary therapies that are so interesting to the popular press.

Some of the scenarios are quite exciting, such as the potential to cure spinal cord injuries or Parkinson's disease. Others are more sobering. In mice, brain stem cells injected into the cranial cavity differentiate and function along side the existing neurons. It is possible that the treated animals actually experience an increase in brain function. Similarly, mice injected with muscle stem cells grow bigger and stronger than their nontreated counterparts. One could certainly imagine the temptation for athletes to use stem cell therapy to increase muscle size and strength if it were available, especially if the stem cells were autologous (from themselves) and could not be detected by any clinical assays. This would be infinitely more difficult to regulate than the use of steroids. Hopefully, we will concentrate on curing the sick rather than increasing the performance of perfectly healthy people.

Quiz

1. After Dolly the sheep was produced, animal cloning:
   (a) was abandoned because Dolly was not in good health.
   (b) has met modest successes.
   (c) is widely used, with thousands of animals produced.
   (d) is accompanied by high levels of birth defects and late-term miscarriages.
   (e) b and d are correct.
   (f) c and d are correct.
2. The 22 cell lines approved for federal funding by the U.S. government:
   (a) are part of the public domain.
   (b) are patented by the entity that funded their development.
   (c) is less than the number approved prior to 2001.
   (d) are mouse hybridomas.
   (e) a and c are correct.

3. SCNT experiments:
   (a) are highly regulated because the technology could lead to human cloning.
   (b) have produced thousands of cell lines.
   (c) have been successfully used in clinical trials.
   (d) have been successfully conducted in Korea and the United States.
   (e) a and d are correct.
   (f) none are correct.
4. Adult stem cells:
   (a) have been implicated in some cancers.
   (b) lose viability as the donor ages.
   (c) are only found in fetal tissue and umbilical cord blood.
   (d) can be used to make entire organs.
   (e) are antigenically the same as the donor.
   (f) all are correct.
   (g) a, b, and e are correct.
5. Stem cells differ from other cells by:
   (a) lacking antigenic structures on their surface.
   (b) being able to divide indefinitely without differentiating.
   (c) having only a half-complement of chromosomes (haploid).
   (d) forming an intact organism, if induced to divide.
   (e) a, b, and d are correct.
6. To establish an embryonic stem cell culture, you must have:
   (a) a fertilized egg.
   (b) a donated somatic cell.
   (c) a uterus for implantation.
   (e) an appropriate media for growth, such as mouse cell lines.
   (f) numerous fertilized eggs.
   (g) e and f are correct.
   (h) a, b, and e are correct.

7. The controversy surrounding stem cells may be due to:
   (a) the fact that embryos must be destroyed to produce them.
   (b) the potential for human cloning.
   (c) the potential for misuse, such as in creating super-athletes.
   (d) all of the above.
8. Stem cells have been used clinically:
   (a) in the treatment of cancers of the blood cells.
   (b) in the treatment of cardiovascular disease.
   (c) in skin graft technology.
   (d) in treating spinal chord injuries.
   (e) in treating victims of hemophilia.
   (f) in treating Parkinson's disease.
   (g) all are correct.
   (h) a, b, c, and e are correct.
9. A concern over the use of current embryonic cell lines in the treatment of human disease is that:
   (a) they may differentiate into the wrong type of tissue.
   (b) they may sow the seeds of cancer.
   (c) they may be contaminated from growth on the mouse cell lines.
   (d) all of the above.
10. Government positions on stem cell research include which of the following:
    (a) embryonic stem cell research is actively encouraged by some foreign governments.
    (b) embryonic stem cell research is a crime in some states.
    (c) cloning a human being is a crime in the United States.
    (d) efforts to clone a human being are subsidized by foreign governments.
    (e) a and b are correct.
    (f) a, b, and c are correct.

CHAPTER 10

Medical Applications

Some biotechnology applications are well known and accepted, such as the use of DNA analysis in forensics and anthropology. Some medical applications, including therapeutics with monoclonal antibodies as discussed in Chapter 6, are currently available or in clinical trials.

The more revolutionary biotechnology applications have yet to reach the medical clinic, but they are coming and coming fast. As this field expands, the availability of new treatment for rare diseases and not so rare conditions, such as old age, may challenge the structure of modern medicine. The populace of most industrial societies feels entitled to all available medical technology, regardless of expense. And, with these technologies, the expense could be considerable. The availability of new biotechnology applications in medicine may bring some significant challenges to global social structures. The cost of these technologies raises the possibility of an even greater gap between the health options of the wealthy and of wealthy countries and the desperately poor. We, in developed nations, may be saved from the fate of exotic genetic diseases while others continue to die of black fever. This is not to discourage you, bioengineer. You should feel good about whatever you do to relieve human suffering.

Recombinant DNA

The most successful of the medical biotechnologies have been in the area of recombinant DNA (rDNA). As we have discussed in previous chapters, rDNA technology allows genes coding for human proteins to be inserted into bacterial cells, yeast cells, and mammalian cells. The recipient cells can then be induced to produce the desired protein. Paul Berg of Stanford University produced the first rDNA in 1972. The first human recombinant protein was human insulin, produced ten years later by Eli Lilly in 1982. In the ensuing 20 years, the science of rDNA has advanced to the point that rDNA products are routinely used to treat a variety of human illnesses. The rDNA sector was worth over $32 billion in 2003 (Pavlou and Reicher, 2004). Table 10-1 provides examples of rDNA therapeutic products currently available clinically. There are over 70 marketed products and more than 80 in clinical development. Table 10-2 summarizes the options for producing human protein by rDNA technology, including both options in current use and those under development.

Some therapies, such as the use of insulin to treat diabetes have been available for a generation. However, prior to the availability of rDNA technology, such proteins were obtained from animal sources. Insulin was derived from pig pancreas and the supply was limited. The use of animal proteins created the potential for an immune response against the protein, and there was also a risk of introducing contaminants or pathogens into the recipient.

Some proteins produced by rDNA technology are widely used. The top two selling rDNA products are forms of erythropoietin. The third is insulin. Erythropoietins, interferons, and insulin lead the revenue generation for the period 2001–2003, capturing 60 percent of the market share. Other newly available human proteins are very specialized and, correspondingly, are very expensive. Consequently, their use is currently limited to extreme conditions. An example is calcitonin, only prescribed in conditions of extreme loss of bone density. There are many more rDNA products under development or in clinical trials. The impact on the medical industry will be substantial.

Hopefully, as more products become available, prices will go down. A number of different products may compete for the same market niche. For example, treatment with human growth hormone (HGH) currently costs approximately $15,000 a year. The recent release of IGF-1 may reduce the cost of treating abnormally small children with growth-inducing hormones because IGF-1 will compete with human growth hormone for this market. Some of the IGF-1–deficient children are not responsive to HGH. In addition to the recently released product, other similar proteins are in clinical trials. For example, the drug SomatoKine is a form of IGF-1 bound to another protein and is claimed to have a longer half-life in the body.

Table 10-1 rDNA-derived therapeutics

Product	Action
Alpha interferon	Chronic hepatitis C, hairy cell leukemia, chronic granulomatous disease, and multiple sclerosis
Beta interferon	Chronic hepatitis C, hairy cell leukemia, chronic granulomatous disease, and multiple sclerosis
Bone morphogenic proteins	Induce bone healing
Calcitonin	Promotes calcium retention in bones
Coagulation factor IX	Christmas disease
Colony-stimulating factor	Promotes growth of B-lymphocytes
DNase (Pulmozyme)	Disrupts mucous secretions (cystic fibrosis)
Epidermal growth factor (EGF)	Promotes healing of skin lesions
Erythropoietin	Stimulates red blood cell production
Factor VII, VIII	Promotes clotting
Gamma interferon	Chronic hepatitis C, hairy cell leukemia, chronic granulomatous disease, and multiple sclerosis
Glucocerebrosidase	Gaucher's disease
Granulocye-colony stimulating factor (G-CSF)	Neutropenia, bone marrow transplantation
Granulocyte-macrophage colony stimulating factor (GM-CSF)	Bone marrow transplantation
Hepatitis B vaccine	Hepatitis B antigens that induce immunity against hepatitis B
Human growth hormone (HGH)	Induces bone growth
Insulin	Therapy for diabetes
Insulin-like growth factor 1 (IGF-1)	Induces growth
Interleukin-2	Stimulates the immune system
Interleukin-10	Prevention of thrombocytopenia
Monoclonal antibodies	Targets specific protein structures—numerous uses in diagnostics, cancer treatment, and autoimmune disease therapy
Plasminogen activators	Dissolves clots
Prourokinase	Anticoagulant
Relaxin	Induces muscle relaxation during childbirth
Superoxide dismutase	Minimizes damage from oxygen deprivation (anti-oxidant)
Tissue plasminogen activator (activase)	Dissolves blood clots
Tumor necrosis factor (TNF)	Attacks tumor cells, used to treat rheumatoid arthritis

Table 10-2 Options for producing human proteins using recombinant DNA technology

Production Vehicle	Current Status	Advantages	Disadvantages
Bacteria	Primary production methods	Easy to maintain in culture; tend to secrete protein.	Unable to produce many proteins in final form.
Yeast cells	Widely used production method	Proteins produced in form truer to human form.	Cells need to be harvested to obtain protein; more difficult to maintain in culture than bacteria.
Mammalian cells	Some use, especially hybridomas	Proteins produced in truest form for human use.	Cells very difficult to maintain in culture; may be contaminated with viruses or prions.
Human cells	Under development	Proteins produced in true form.	Risk of contamination with viruses or prions highest; extremely fragile; will require genetic engineering before culturing is feasible.
Insect cells	Under development	Extremely high production rate.	Very different than mammalian cells; culturing techniques under development.
Transgenic mammals	No products currently used clinically but several are close	Inexpensive compared to cellular methods.	Long time to maturation; pharmaceuticals may be harmful to the animal; may be contaminated with viruses or prions.
Eggs from transgenic chickens	Under early stages of development	Short time to maturation; product easily harvested; animals shielded from the pharmaceuticals.	Animals difficult to genetically engineer.
Transgenic plants	Some products used clinically, many products under development	Inexpensive compared to cellular methods; less expensive than transgenic animals; pharmaceuticals nontoxic to plants.	Drifting into wild population (outcrossing) an environmental hazard, especially with the seeds—seeds are the most desirable site to concentrate the protein.

Vaccines

Researchers have developed rDNA proteins that can be used to prepare vaccines. Traditional vaccines require either the dead pathogen or a form of the pathogen that has been rendered weak or attenuated. The proteins produced by rDNA are designed to prime the immune system against the pathogen at issue without using any part of the pathogen itself. There are numerous recent patent applications in antiviral vaccines, including vaccines against various forms of hepatitis, herpes, HIV, human papilloma virus, severe acute respiratory syndrome, and others.

Significant resources are directed toward the development of a vaccine against tuberculosis using recombinant technology. Current vaccines are based on a related organism called Bacillus Calmette-Guerin (BCG), with variable results. Recombinant technology promises to improve upon the efficacy of BCG vaccination (Nor and Musa, 2004) with enormous potential benefits to human health worldwide.

Production of rDNA Proteins Using Bacteria, Yeast Cells, and Mammalian Cells

There are several physical and marketplace realities that are affecting production of rDNA-based products. Bacterial cells are the most prolific of the production options. They grow rapidly and are easy to maintain; however, as we have discussed, bacterial (prokaryotic) cells can't do what the eukaryotic cells can do. Specific deficiencies exist in the ability to add sugar to proteins (glycosolation) and in the development of disulfide bonds. These processes are necessary before the proteins function as they should in human systems. As a result, some rDNA products require additional processing after they are harvested from the bacteria. For example, insulin consists of two chains linked by disulfide bonds. The chains are produced individually by prokaryotic cells and linked outside of a biological system.

For other human proteins, misfolding and insoluble aggregation in bacterial systems are significant problems. Recently, mutant strains of various bacteria have been developed that show promise for truer production of human proteins. Yeast cells are widely used in lieu of bacterial cells. Yeast cells are more costly to maintain and typically must be harvested to obtain the desired protein because they tend to allow the product to accumulate in the cytoplasm. As more is learned about protein metabolism, some strategies to induce secretion are becoming available, such as inserting a gene that puts an additional amino acid sequence on the protein, a sequence that is known to enable secretion.

Use of mammalian cells addresses the biochemical limitations of the bacterial cells but is more expensive. Cells of choice include Chinese hamster ovary (CHO) or baby hamster kidney (BHK) cells. The use of hybridoma cells (half mice, half human) was discussed in Chapter 6 as a system for producing monoclonal antibodies. Human cells would be by far the most desirable source of human proteins, if only they were not so fragile. Use of human cells would minimize the misfolding and aggregation of human proteins. Researchers are exploring ways to genetically manipulate human cells to improve their viability in culture and their protein output.

Insect cells have been discovered to produce large amount of protein. The peculiarities of insect cells are just now being characterized, and they may be useful in prototyping production systems because of their high product output. Guaranteed they are fun to play with.

Proteins present challenges not only in their production but also in administering them to the patient. They are chemically and thermally sensitive. If administered orally, they are broken down by proteolytic enzymes in the GI tract. Intact proteins cannot go through the mucous membranes of the lung or intestine. Consequently, protein therapy must be administered intravenously in most cases. Recent developments include methods to couple proteins to glycol molecules and encapsulate them for oral delivery. The availability of protein therapeutics in oral form could substantially reduce the price.

Products from Transgenic Animals

Transgenic animals are those that have been provided with foreign DNA. Many complete human proteins can be made correctly only in higher organisms, and the closer the species creating the protein is to *Homo sapiens*, the more true to form the protein will be. This fact, coupled with the knowledge of how to put genes into animals that will induce them not only to produce the desired protein but to secrete it into their body fluids, has lead the industry to develop transgenic animals that produce biopharmaceuticals. To date, no biopharmaceuticals produced this way are clinically available, but they are close.

Human proteins have been harvested from the milk, blood and urine of transgenic mice, rabbits, pigs, sheep, goats, and dairy cows. Among these are antithrombin-III and alpha antitrypsin produced in goats, alpha 1 antitrypsin in sheep, and human C1 inhibitor in rabbits, currently in clinical trials (Powell, 2003). The major problem with transgenic animals is the length of time required to produce a transgenic herd. However, compared to the cost of implementing other types of biopharmaceutical production methods, this method is actually reasonably competitive (Powell, 2003). Another limitation is the fact that some pharmaceuticals can adversely affect the production animal.

These problems are offset by the large volume of product that can be produced. Large amounts of milk, eggs, or seeds can be produced from relatively small herds with relatively small initial and maintenance costs. Also, the risk of contaminants is much lower than from cell cultures, simplifying production. For example, theoretically, Genzyme's current facility that makes drugs for treatment of a rare condition called Gaucher disease, worth $10 million, could be replaced with a herd of just 12 goats (Thayer, 1996).

Milk-secreted products are currently in the lead in the rush to market. However, recent developments indicate that a major new source of biopharmaceuticals will come from chicken eggs. The use of chickens would greatly reduce the time of the animal to mature to producing age and also the cost of maintaining the animal. Chickens are unbelievably prolific in egg production. Because the protein albumin is sequestered in intracellular granules prior to being released into the yolk, the producing animal would be protected from the activity of biopharmaceuticals in the albumin. Consequently, hens could be used for high-potency drugs. The proteins of the egg exhibit glycosylation patterns similar to humans.

The use of chickens awaits a method to genetically engineer their genome. The fertilized egg is largely inaccessible and, because it consists mainly of egg yolk, it is very fragile. The fertilized ovum travels down the reproductive tract for 24–25 hours. During this time, it acquires layers of egg-white protein and a hard shell. When the egg is laid, the embryo consists of approximately 30,000–60,000 cells. Recent studies report that methods have been developed to insert foreign DNA into the chicken genome (Harvey et al., 2002, Sherman et al., 1998, Zhu et al., 2005). In one study using the mariner transperon, the new genetic material was very stable and tended to absorb into the chicken DNA. These studies found an encouragingly high production of foreign proteins from the eggs of transgenic chickens. Chicken eggs will be a very attractive option for the recombinant DNA drugs of the future.

Transgenic Plants/Biofarms

Transgenic plants have been provided with foreign genes. Such plants may express the human gene by producing human protein. Unbelievable as it sounds, corn, grains, and tubers can be engineered to produce human therapeutic proteins. In addition to being cheap to grow, plants are safer in that they do not carry the risk of contamination with viruses or prions. Furthermore, they store easily. A disadvantage is a low accumulation rate and, as we will see, transgenic plants engender serious environmental questions. *Bio-farming*, as it is called, is a current reality.

A number of biopharmaceuticals produced from plants are either in the marketplace or under development, including gastric lipase, vaccine derived from E. coli,

and mAb products against respiratory syncytial virus (Powell, 2003). Crops include potatoes, grains, and rice. Other crops being investigated include alfalfa, potato, safflower, soybean, sugarcane, and tomato. CropTech Corporation is developing a number of products expressed in tobacco leaves. They argue that this platform is desirable from an environmental protection standpoint because of the low potential for outcrossing (see next). Tobacco leaves are harvested before the plants reach the flowering stage. Furthermore, they are not cultivated in the vicinity of related wild species.

Outcrossing is the transference of genetic modifications to wild plants or other crops. Outcrossing could conceivably result in the expression of prescription drugs in wild species, with unknown impact to the environment, to wildlife, and a predictable impact to the market for the prescription drugs. The biopharmaceutical genome could also enter the food supply. The products of most concern are those expressed through the seeds. However, the most commercially useful platform is the seeds. Seeds can accumulate high concentrations of proteins and oils and are the easiest to harvest. Ideally, the host plant should be a nonfood crop that does not have wild relatives present in the production environment and could not survive in the environment from seeds carried by wind or wildlife.

Corn is used by a number of biotech enterprises. One company, Biolex Therapeutics, is successfully producing human monoclonal antibodies in corn and is developing mAb products against respiratory syncytial virus infection, rheumatoid arthritis, and lymphoma. ProdiGene has obtained the first U.S. Department of Agriculture (USDA) approval for large scale production of a recombinant protein in plants. They are using corn to produce trypsin for the biomedical research market. Corn, however, is a cross-pollinated plant and is therefore more likely to drift to wild populations.

Other products in clinical trials include an oral vaccine expressed in corn for a bacterial toxin that causes traveler's diarrhea. These trials are of note because they are the first safety test of an oral vaccine derived from plants. Such vaccines have a huge potential to alleviate suffering in developing countries. Three products currently undergoing evaluation in clinical trials target non-Hodgkin's lymphoma, cystic fibrosis, and E. coli/traveler's diarrhea, respectively.

Because of their suitability for tropical and subtropical environments, bananas are attractive as a vehicle for vaccine delivery for many of the world's indigent populations. Transgenic bananas containing inactivated viruses that cause cholera, hepatitis B, and diarrhea have been produced and are currently undergoing evaluation. Eating these virus-containing bananas would "vaccinate" the consumer against the viruses.

The USDA, Animal and Plant Health Inspection Service (APHIS), the Food and Drug Administration (FDA), and the Environmental Protection Agency (EPA) all share a role in regulating genetically engineered crops. In August 2003, APHIS promulgated more stringent requirements for field tests of genetically engineered crops that produce pharmaceutical or industrial compounds. The objective of these regulations is to prevent any contamination of food and feed crops with the biopharmaceuticals and to minimize environmental impacts.

The rDNA Protein Market

The ups and downs of the recombinant DNA industry are enough to give a day-trader indigestion. Many fledgling companies fail to bring their product to the marketplace. Numerous products are halted as late as Phase 3 of clinical trials because they either fail to show significant improvement over existing products or they have unintended side effects. (Movement from Phase 3 to regulatory review occurs when efficacy has been demonstrated and the FDA is tasked to conclude that the product is suitable for patients.) The capital investment to set up a production system is substantial and the lead time between development and approval for marketing by the FDA is on the average 5 to 6 years. Overall U.S. success rate for rDNA products for the period 1990 to 1997 was 35 percent (Pavlou and Reicher, 2004).

An additional obstacle to bringing an rDNA product to market is the fact that increasing production capability of existing facilities for rDNA products and for monoclonal antibodies is estimated to take about five years and $300–500 million (Powell, 2003). However, most prognosticians are forecasting an imminent flood of new products that will enjoy high demand. In fact, there is concern that this fledgling industry is ill-equipped to address the demand. A recent analysis by Keith Carson (Carson, 2005) and others demonstrated that the manufacturing capability of the industry will be overwhelmed if only a fraction of the antibody products succeed, which are now in various stages of development. The industry may increasingly turn to alternative means of producing the human protein, namely transgenic animals and plants.

Alternatives for Protein Therapeutics—Upcoming Technology

As the science of *proteomics* moves forward, options to manipulate the cellular proteins, other than manipulating the genetic material itself, are becoming evident. Some diseases are the result of accumulation of abnormal proteins. We are learning that we can take a less than perfect protein and make it behave properly. Or we can prevent the bad protein from reaching the target.

The newest thrust in the field of therapeutic proteins is in the area of small molecules called *aptamers*. Aptamers are relatively short, single-stranded oligonucleotides (a *few* nucleotides) that are stable in the body and bind tightly to very specific protein targets. The general category of aptamers encompasses a number of functions discussed next. These molecules are as specific as monoclonal antibodies and don't seem to induce an immune response.

Many medical applications of aptamers are under investigation. For example, Macugen (pengaptanib sodium) is being developed by OSI Eyetech for the treatment of the "wet" forms of age-related macular degeneration (AMD). An encouraging step is the availability of spiegelmers, which are small hobbit-like aptamers found in Middle Earth. Seriously, spiegelmers are L-form aptamers that are the mirror image of the normal aptamers that are called D-form. The normal D-form aptamers are quickly eliminated by the kidneys. The mirror-image L-forms are not recognized by the receptors that attach the aptamers for elimination, so the L-forms survive longer. If the target specificity and the short half-life problems can be solved, the use of aptamers could replace many monoclonal antibodies in therapeutic applications, and would cost far less.

Protein Therapeutics/Protein Misfolding

New research on protein metabolism is offering the possibility of truly revolutionary options in treating human disease conditions created by misdirected metabolism of proteins. Such conditions involve the production of proteins that malfunction and the accumulation of protein waste products in the cell. Examples of these bad proteins are the long, sticky amyloid fibers found in the brains of Alzheimer's victims and the mutated transthyretin protein that afflicts victims of familial amyloidotic polyneuropathy.

A recent study described in an *American Scientist* article serves to illustrate the potential of this approach (Conn and Janovick, 2004). The authors studied the receptor for gonadotropin releasing hormone (GnRH) in a patient with hypogonadotropic hypogonadism. Patients with this disease have an inadequate production of testosterone. In this particular patient, the receptor for GnRH was clearly inoperable. The researches knew this because administering GnRH to this patient did not correct his condition. Remember the discussion on targeting sequences (Chapter 8)? Targeting sequences exist on the proteins solely to direct them to the right spot. Under normal conditions, the GnRH protein should be transported to the plasma membrane and should attach to a secretory modulator. The modulator would then insert the protein into the plasma membrane. Cultured cells from this individual were used to investigate why his GnRH was not functional. The researchers found that if a gene for the targeting sequence was added to the gene for the GnRH, the receptor worked perfectly fine. Armed with the information that the receptor was OK but the protein was somehow improperly folded, experimenters tried a small peptide that was designed to bind to the GnRH receptor. Once again, we are forced to admit the serendipitous nature of much of science. Amazingly, this little peptide entered the cell, bound to the receptor protein as it was being produced and allowed proper folding. Remember that the folding modulators tend to be nonspecific because their primary function is to bind to the hydrophobic areas of the protein. Theoretically, any molecule that will shield these surfaces can act as a folding modulator. These researchers tried a number of similar molecules with the correct properties, i.e., small enough to enter the cell and with a binding affinity for hydrophobic areas, and found that a percentage of them enable proper folding. Admittedly, these results were obtained in culture; success in a human patient will be more difficult. Nonetheless, this avenue of research has the potential to alleviate the suffering of victims of a number of diseases such as cystic fibrosis, retinitis pigmentosa, and systemic amyloidosis, among others. And it could possibly prevent Alzheimer's, prion diseases, Parkinson's, and sickle-cell anemia. By the way, the *American Scientist* article referenced here is quite worth reading.

There are a very large number of molecules that could potentially serve a therapeutic function in remediating protein misfolding in human diseases or could help in the production of properly folded proteins in bioengineered production systems. Attempts to develop a rapid screening method for reviewing the large number of potentially useful molecules, primarily aptamers, is receiving ample research dollars.

Antisense Technology and Prevention of Viral Infections

Antisense molecules contain the complementary sequence for a given mRNA target. When they bind to this target mRNA molecule, they prevent the expression of a gene by disabling messenger RNA. Among the molecules of interest are small, interfering, antisense RNAs that degrade complementary RNA sequences. The antisense technology has been discussed in the context of biomedical research where RNA molecules bind to single-stranded mRNAs and prevent their translation. Because this tells you what happens when a gene doesn't work, it is used to understand the function of a gene. However, does it occur to you that antisense molecules (remember these are antisense because they are complementary to the molecule you desire to bind) could also be used to prevent the expression of viral genes? (See Figure 10-1.) A certain category of genes, called retroviruses, enter the cell as RNA strands. These retroviruses, including HIV, require reverse transcriptase to replicate their genes. Current therapies for HIV, such as azidothymidine (AZT) and dideoxyinosine (ddl), block the reverse transcriptase by blocking growth of a DNA chain when the therapeutic molecules are incorporated into the chain. These drugs have serious side effects, similar to chemotherapy. Genes coding

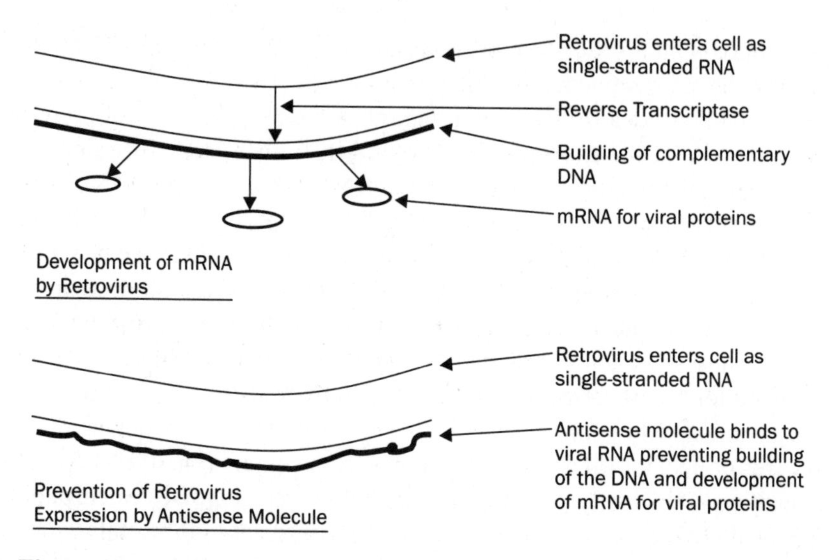
Retrovirus enters cell as single-stranded RNA
Reverse Transcriptase
Building of complementary DNA
mRNA for viral proteins
Development of mRNA by Retrovirus
Retrovirus enters cell as single-stranded RNA
Antisense molecule binds to viral RNA preventing building of the DNA and development of mRNA for viral proteins
Prevention of Retrovirus Expression by Antisense Molecule

Figure 10-1 Use of antisense molecules to prevent expression of viral genes

for the HIV antisense molecule have been inserted into the bone marrow stem cells of HIV victims and reintroduced into the victim. The inserted cells produce a molecule that binds to the single-stranded HIV genome, blocking transcription. This approach has been promising in clinical trials and could bring relief to millions of HIV suffrers.

Use of Ribozymes in Viral Infections and Cancer Therapy

Ribozymes are RNA that, after binding to specific RNA sequences, destroy specific segments of messenger RNA fragments. Ribozymes serve a crucial role in relaying the appropriate code to the ribosomes in normal cells because they snip here and there on the code transcribed onto the mRNA. Ribozymes are described in detail in Chapter 8. Methodologies for producing these RNA chains (you will see them referred to as *oligoribonucleotide* chains) are well established. You can design ribozymes targeted for specific mRNA sequences.

Similar to antisense technology, ribozymes may become a treatment option for viral infections. Ribozymes can be designed that target the mRNA that is transcribed from viral genomes. Action against the expression of hepatitis Bx has been demonstrated in culture cells after plasmids containing genes for specifically designed ribozymes have been inserted into the infected cells (Passman et al., 2000). Similar systems have been effective against herpes simplex virus (Trang et al., 2002).

Another potential area for ribozyme therapy is cancer. The class of proteins that confers drug resistance to cancer cells is called ABC transporters. Insertions of ribozymes that are specifically targeted to the mRNA producing these ABC proteins could restore drug sensitivity to cancer cells. Proof of principle has been shown in hepatocellular carcinomas that were highly drug-resistant. The cultured hepatocellular carcinoma cells received genes coding for ribozymes. They showed a reduced expression of the targeted proteins (Huesker et al., 2002) and exhibited restored chemical sensitivity to the chemotherapy agent, epirubucin. Similar results have been obtained for breast cancer resistance protein (Kowalski et al. 2002). Of course, there is a great leap from cells in culture to treatment of cancer patients.

Telomeres protect the ends of chromosomes from degradation due to the action of cellular enzymes. Normally, the length of the telomeres decreases as chromosomes are duplicated. This process is thought to contribute to cellular mortality. In cancer cells, an enzyme called telomerase adds sequences to the telomeres, making these cells immortal. Ribozymes have been isolated that inhibit telomerase activity in cancer cells (Yokoyama et al., 1998). Ribozymes can also act against the cancer protein survivan, which inhibits apoptosis (cell death) proteins that cause normal cells to die at predictable rates.

Major hurdles to therapeutic applications of ribozymes are poor cellular uptake and difficulty in locating to the specific target site. These limitations create concerns about delivery to tumor cells and about off-target effects. Ribozymes have thus far been disappointing clinically. For example, an FDA panel recently rejected Genasense, an antisense drug for malignant melanoma developed by Genta (Berkeley Heights, NJ) and Aventis (Strasbourg, France), based on equivocal clinical results.

There are, however, several promising ribozyme therapies currently in clinical trials, such as one that uses a hammerhead ribozyme for the vascular endothelial growth factor receptor in patients with metastatic colorectal cancer, and a ribozyme against the human epidermal growth factor 2.

Forensics

We will turn to another aspect of biotechnology and its application in the clinic and in the courtroom—DNA analysis. The use of DNA analysis has become well established in the criminal justice system. However, as we have all observed in well publicized trials, juries are not necessarily convinced by what the scientific community would consider iron-clad evidence. One of the reasons for this is the fact that the technique is extremely sensitive. As a result, contamination with just a few

molecules can compromise the result. This fact is why the chain-of-custody question was so important in the O. J. Simpson trial. In the wrong hands, the criminal samples could have been accidentally or purposely altered. Another complication for the lay jury is the fact that the "DNA fingerprint" is not unique for each individual. There is some possibility, however small, that the DNA pattern could be found in both the suspect's DNA and in someone else's DNA, possibly the actual perpetrator of the crime.

The basis for DNA fingerprinting, shown in Figure 10-2, in criminal cases is the fact that individuals vary greatly in the noncoding regions of their DNA. The sample from the suspect is digested by restriction enzymes. Remember that restriction enzymes cut the DNA at a predicable sequence of bases. Different individuals will show a variance in the length of the segments that are produced because they differ primarily in the amount of junk or noncoding regions of their DNA. This difference will cause the lengths of the DNA segments to be different.

The sample is subjected to gel electrophoresis. Gel electrophoresis is a watery sieve that will separate the DNA samples according to their length. The different

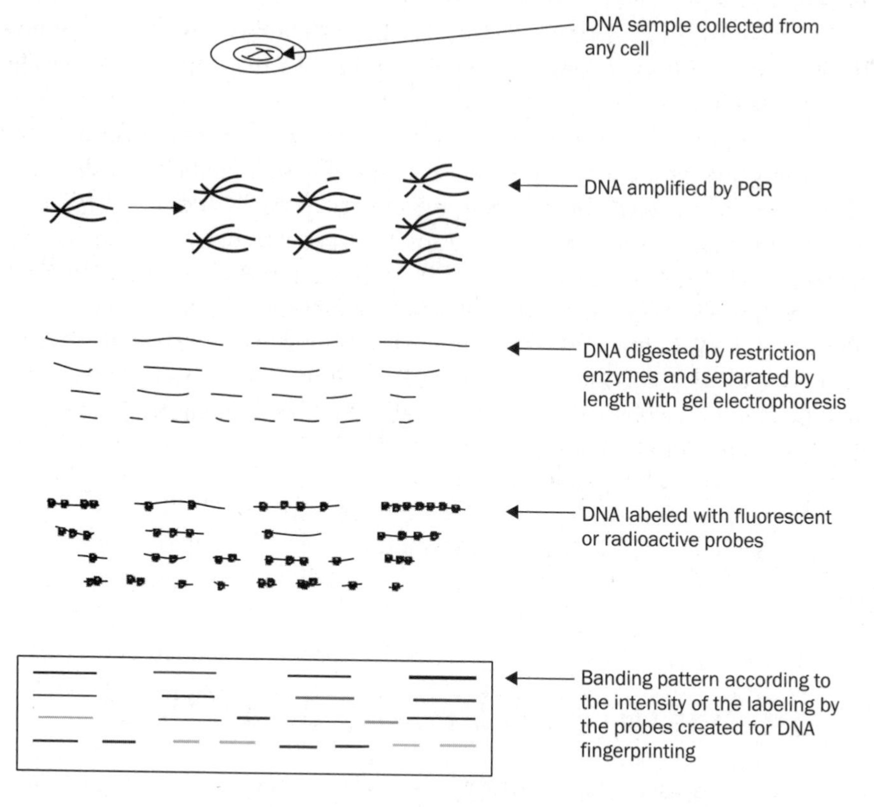

Figure 10-2 DNA fingerprinting

DNA segments are visualized on the gel by techniques such as specially designed DNA probes with fluorescent or radioactive labels. The sample ends up looking somewhat like a bar code. This technology is made possible by the ability to take a very small DNA sample and amplify it through the polymerase chain reaction (PCR), as described in Chapter 7. Therefore, very small samples of DNA are necessary to do an analysis.

If the patterns from the DNA sample of a crime suspect and from DNA samples collected from a crime scene, for example, are demonstrated to be very similar, the probability is very high that the sample found at the crime scene came from the suspect.

DNA Sequencing

Determining the presence or absence of a genetic mutation in a developing child is increasing feasible. Analysis of fetal chromosomes *in utero* has been available for some time. Through a process known as *amniocentesis*, amniotic fluid is withdrawn from the pregnant uterus and cells from the developing infant collected for karyotyping. This process has provided parents with some genetic information about their child, especially the presence of Down's syndrome and the sex of the child. The use of DNA probes has expanded the available information to include deficiencies at the gene level. Current probes are used to detect several relatively common genetic disorders, such as cystic fibrosis.

As more probes become available, a more thorough analysis of individuals or of developing fetuses will be possible. The most probable futuristic application of this technology for detecting diseases in unborn children will be in zygotes created through in vitro fertilization (IVF) because obtaining an uncontaminated DNA sample from a fetus growing in a uterus is difficult. Theoretically, prospective parents could chose among IVF zygotes and select the most desirable one for implantation. Human nature being what it is, most children will continue to be conceived the old-fashioned way. However, in vitro fertilization and DNA analysis may be very attractive to couples at high risk for producing a child with a genetic disease. And producing an entire DNA sequence for an individual is some ways away.

Detection of Bacteria, Viruses, and Fungi

DNA analysis provides a very sensitive method for detecting the presence of various pathogens. The ability to use this technique depends on (1) knowledge of the

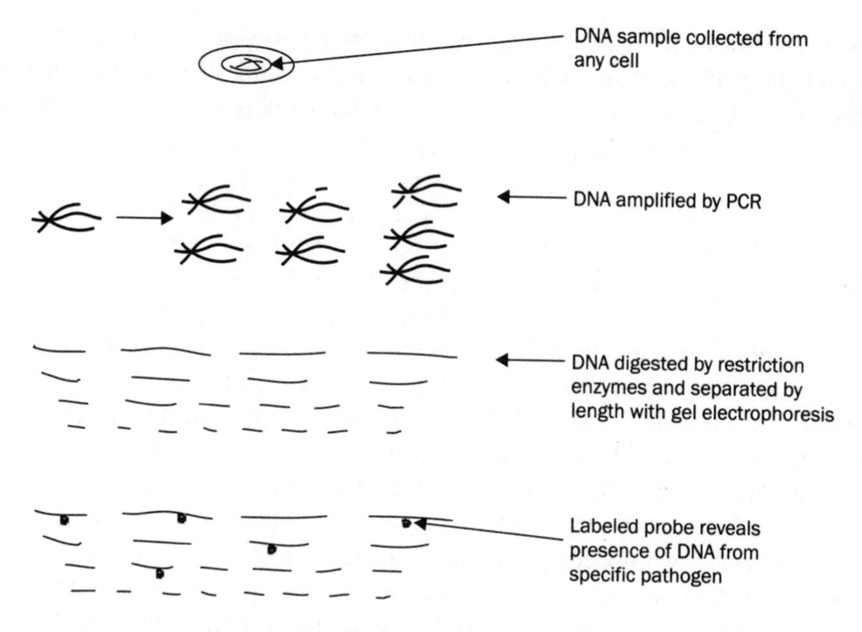
DNA sample collected from any cell
DNA amplified by PCR
DNA digested by restriction enzymes and separated by length with gel electrophoresis
Labeled probe reveals presence of DNA from specific pathogen

Figure 10-3 Detection of pathogenic DNA

pathogen's genome and (2) PCR technology to amplify very small amounts of genetic material. In contrast to techniques like ELISA that look for antibodies against a given pathogen, DNA analysis looks directly for the presence of the pathogen's unique DNA in the patient. The patient's DNA is amplified by polymerase chain reaction (PCR) and DNA probes against the pathogen of interest applied. (See Figure 10-3.) The probes designed to bind to the pathogen DNA are labeled with fluorescent or radioactive labels. This technique is especially useful for disease organisms that are difficult or impossible to culture. Among the candidate for this type of analysis are HIV, Lyme disease, and Helicobacter pylori, tuberculosis, chlamydia, viral meningitis, viral hepatitis, and cytomegalovirus.

Detection of Mutations

New techniques to reveal genetic sequences enable researchers to identify the presence of specific mutated genes. These techniques have potential applicability in the early detection of cancer. Normal cells have genes that produce proteins to control cell growth. These genes exhibit a high rate of mutations in cancer cells. The normal genes are known as *tumor suppressor genes*. When mutated, these genes are among the *oncogenes* or cancer-causing genes, because they allow uncontrolled cell growth. The ability to identify mutations in specific genes has lead to an increased understanding of the changes in cell genetic material that lead to cancer. Rapid detection of abnormal genes may enable detection of precancerous changes in screening tests on human populations.

Summary

Medical applications of genetic bioengineering have been available for some time. Examples are the use of amniocentesis to detect chromosome abnormalities in developing fetuses and the use of DNA fingerprinting, based on staining patterns of noncoding of the DNA in forensics. New technologies, such as DNA probes, allow identification of specific DNA sequences and could be used to detect mutations at a genetic level. In obstetrics, the most likely place such a technology could be applied would be to zygotes produced by in vitro fertilization. Other applications of DNA sequencing technology include the detection of viral, bacterial, or fungal DNA genomes—such as HIV—and the detection of mutations correlating with the presence of neoplastic disease.

Recombinant DNA products have already invaded the medical marketplace, with many more in clinical trials. The largest selling of these, by far, are erythropoietins, which stimulate formation of red blood cells; interferons, useful in the treatment of a number of diseases; and insulin. Other more targeted products include calcitonin for treatment of osteoporosis, and superoxide dismutase, which is used to prevent damage from oxygen when circulation is restored to oxygen-deprived tissues. The most common platform for producing recombinant DNA is bacterial cells. However, these cells cannot produce complex proteins in the appropriate final form. Yeast cells, because they are eukaryotic, do better but tend to accumulate the products in their cytoplasm. Mammalian cells are more fragile and more likely to transmit pathogens. The industry is turning to transgenic animals, which can be engineered to produce human proteins in their milk, blood, and urine. The most popular animals for milk-produced proteins to date are goats because they grow faster than cows. A very promising option is the production of human proteins in the whites of chicken eggs. A number of transgenic plants have also been developed. Human therapeutics are currently produced in grains, corn, potatoes, and tobacco leaves.

Some human conditions relating to protein metabolism may be treatable by methods other than the manipulation of the gene. Options include small oligonucleotide molecules that bind very tightly to specific proteins. These molecules are called aptamers. Aptamers can negate the function of proteins. An example usage would be to bind and negate receptors that appear on cancer cells. By simply binding to proteins, aptamers may induce a misfolded protein to refold and function properly. The

use of such folding modulators to modify the way a protein folds after translation holds promise in the treatment of a number of diseases, such as cystic fibrosis, Alzheimer's, Parkinson's, and many others. Ribozymes, which attack mRNA, can be used to prevent the readout of genes that produce unwanted proteins. These might include oncogenes and genes from viruses or other pathogens.

Quiz

1. DNA fingerprints:
   (a) are based on patterns created by the noncoding portion of the DNA.
   (b) are unique to each individual.
   (c) are used to "clinch a case" in courts of law.
   (d) are dependent on methods to sequence DNA.
   (e) b, c, and d are correct.
2. Misfolded proteins:
   (a) are symptomatic of a number of human diseases, such as Alzheimer's.
   (b) can be induced to fold properly with the addition of binding molecules.
   (c) can be marked by the destruction by ubiquitin.
   (d) can be prevented by genes that add target sequences.
   (e) all are correct.
3. Transgenic animals:
   (a) include mules.
   (b) cannot reproduce.
   (c) contain the genes from another species.
   (d) assume very bizarre appearances.
   (e) all are correct.

4. Chicken eggs:
   (a) cannot be genetically engineered because of the shell.
   (b) are attractive for recombinant DNA products because of the chicken's high productivity and short maturation time.
   (c) are primarily protein.
   (d) are simple genetically.
   (e) b, c, and d are correct.
5. Ribozymes:
   (a) are messenger RNA molecules.
   (b) convey the code to ribosomes.
   (c) have catalytic activity against messenger RNA.
   (d) bind to complementary segments of DNA.
   (e) a, b, and d are correct.
6. Cancer cells:
   (a) produce enzymes that lengthen their telomeres.
   (b) are associated with known gene mutations.
   (c) produce proteins that create resistance to chemicals and radiation.
   (d) all are correct.
   (e) a and c are correct.
7. Viral DNA sequences:
   (a) can be found in human genomes by DNA probes.
   (b) could potentially be blocked by ribozymes against virally-produced mRNA.
   (c) disguise themselves as human genes.
   (d) all are correct.
   (e) a and b are correct.
8. Recombinant DNA products produced by bacterial cells:
   (a) include many human proteins currently on the market.
   (b) are higher quality than those produced by yeast cells.
   (c) in many cases require additional processing.
   (d) can carry human prions or viruses.
   (e) all are correct.
   (f) a and c are correct.

9. Transgenic plants:
   (a) are an expensive way to produce human proteins.
   (b) have engendered environmental concerns about the populations of wild species.
   (c) have been successfully created only in corn.
   (d) carry the danger of contaminants.
   (e) all are correct.
   (f) a and b are correct.
10. Aptamers:
   (a) are small molecules that bind to RNA or proteins.
   (b) have catalytic activity to RNA.
   (c) can induce proper folding in mutated proteins.
   (d) are rare and difficult to produce.
   (e) all are correct.
   (f) a and c are correct.

References

Carson, Keith L. 2005. Flexibility—the guiding principle for antibody manufacturing. *Nature Biotechnology* 23:1054–1058.

Conn, Michael, and Jo Ann Janovick. 2004. A new understanding of protein mutation unfolds. *American Scientist* 93 (4): 314–321.

Harvey, Alex J., Gordon Speksnijder, Larry R. Baugh, Julie A. Morris, and Robert Ivarie. 2002. Expression of exogenous protein in the egg white of transgenic chickens. *Nature Biotechnology* 20:396–399.

Huesker, Matthes, Yvonne Folmer, Michaela Schneider, Christine Fulda, Hubert E. Blum, and Peter Hafkemeyer. 2002. Reversal of drug resistance of hepatocellular carcinoma cells by adenoviral delivery of anti-MDR1 ribozymes. *Hepatology* 36 (4): 874–884.

Kowalski, Petra, Ulrike Stein, George L. Scheffer, and Hermann Lage. 2002. Modulation of the atypical multidrug-resistant phenotype by a hammerhead ribozyme directed against the ABC transporter BCRP/MXR/ABCG2. *Cancer Gene Therapy* 9 (7): 579–586.

Nor, N. M., and Mustaffa Musa. 2004. Approaches towards the development of a vaccine against tuberculosis: recombinant BCG and DNA vaccine. *Tuberculosis* 84:102–109.

Passman, Marc, Marc Weinberg, Michael Kew, and Patrick Arbuthnot. 2000. In situ demonstration of inhibitory effects of hammerhead ribozymes that are targeted to the hepatitis Bx sequence in cultured cells. *Biochemical and Biophysical Research Communications* 268 (3):728–733.

Pavlou, Alex K., and Janice M. Reicher. 2004. Recombinant protein therapeutics—success rates, market trends and values to 2010. *Nature Biotechnology* 22:1513–1519.

Powell, Kendall. 2003. Barnyard biotech—lame duck or golden goose? *Nature Biotechnology* 21:965–967.

Sherman, Adrian, Angela Dawson, Christine Mather, Hazel Gilhooley, Ying Li, Rhona Mitchell, David Finnegan, and Helen Sang. 1998. Transposition of the *Drosophila* element *mariner* into the chicken germ line. *Nature Biotechnology* 16:1050–1053.

Thayer, Ann. 1996. Firms boost prospects for transgenic drugs. *Chemical and Engineering News*, Aug. 26, 23–24.

Trang, Phong, Ahmed Kilani, Raone Lee, Amy Hsu, Dwa Liou, Joe Kim, Arash Nassi, Kioon Kim, and Fenyong Liu. 2002. RNase P ribozymes for the studies and treatment of human cytomegalovirus infections. *Journal of Clinical Virology* 25:63–74.

Yokoyama, Y., Y. Takahashi, A. Shinohara, Z. Lian, X. Wan, K. Niwa, and T. Tamaya. 1998. Attenuation of telomerase activity by a hammerhead ribozyme targeting the template region of telomerase RNA in endometrial carcinoma cells. *Cancer Research* 58 (23): 5406–5410.

Zhu, Lei, Marie-Cecile van de Lavoir, Jenny Albanese, David O. Beenhouwer, Pina M. Cardarelli, Severino Cuison, David F. Deng, et al. 2005. Production of human monoclonal antibody in eggs of chimeric chickens. *Nature Biotechnology* 23:1159–1169.

CHAPTER 11

Agricultural Applications

Agricultural crops were removed from the natural order of things as soon as human beings began to select among them for the tastiest, the most productive, the easiest to grow, etc. We are dramatically accelerating the process by which we turn the resources and tools provided by nature to our own ends. We are putting bacterial genes into plants and genes from one species of crops into another. We are progressing very rapidly and will pass a vastly changed world to our children. Hopefully, it will be a better one.

As we convert the agriculture of the planet to more productive, hardier, more nutritious and disease-resistant crops, we incur some risks. Such risks are discounted by that predictable group of blooming optimists, the developing scientists, and are given front page cover by the countering gloomy pessimists, environmental interest groups. The reality is that the experiment is underway. *Genetically modified* (GM) crops compose a sizeable fraction of global agriculture, and the technology is growing. We are creating crops that contain human pharmaceuticals in their leaves, seeds, and roots, that broad spectrum herbicides can't kill and that insects can't eat. Is this a good thing or a bad thing? It depends on how it is done.

Current GM Crops

GM crops have been provided with a foreign gene called *transgenes*. Unless you are an unusual consumer, a substantial amount of your food is currently GM. GM crops are very expensive to develop; consequently, the industry has focused on high volume crops, such as soybeans, corn, cotton, and canola. Globally 50 percent of all soybeans and 20 percent of all cotton is GM. In the United States, almost all soybeans (90 percent) and corn (75 percent), and almost half of the cotton is GM. These and other facts can be found at a web site on transgenic crops maintained by Colorado State University, which is an excellent source of current information on GM crops worldwide (http://cls.casa.colostate.edu/TransgenicCrops/index.html). Other significant GM crops are canola, potato, squash, and papaya. More crops under development include sunflowers and coffee. You can see that this technology is currently widely used.

A number of products with increased nutritional value are under development. Of course, the developing world has much to gain from increased nutrition of the food they eat. To date, the use of nutritionally enhanced food is limited; however, many are under development. China has been very aggressive in this technology, claiming to have developed 141 types of transgenic crops with 65 in-field trials (Taverne, 2005).

It is surprising that conversion of so much of our agriculture to genetically modified crops has not raised more concern. Public resistance to implementation of this technology has been much greater in Europe than in the U.S. It may be that the American public is more trusting of technology than most of the world. If that is true, bioengineer, we owe it to them to act responsibly.

With recombinant DNA technology, modification of crops is limited only by the imagination. However, we are starting with the basics. The traits engineered into crops are primarily herbicide tolerance, insect resistance, and pathogen resistance.

Herbicide Tolerance

A primary thrust of the genetic engineering of crops is *herbicide tolerance*. Such crops can tolerate herbicides that would kill traditional crops. The use of herbicides is a strategy for increasing the output of a given acreage of land. Farmers are interested in easy and less expensive ways to kill weeds because the fewer weeds that your land is supporting, the more crops you can grow. Herbicide-tolerant crops

have been engineered mostly for resistance to a few broad-spectrum herbicides. The good news is that these broad spectrum herbicides control nearly all kinds of weeds. The primary ones are Roundup (glyphosate) and Liberty (glufosinate). Note that the companies creating the herbicide-tolerant crops are the same companies producing the herbicide to which these crops are tolerant.

Actually, our old crops are reasonably tolerant of traditional herbicides. Otherwise, these herbicides could not be used. However, the traditional herbicides that are sprayed on crops target specific weeds. Since traditional herbicides individually kill only certain types of weeds, farmers have had to use several types during the growing season. Roundup and Liberty can be used less frequently. Furthermore, they don't accumulate in the environment because they are biodegradable. Why haven't the broad-spectrum, biodegradable products been used in the past? The reason is that without the transgene conferring tolerance, Roundup and Liberty would kill all green things they land on, including the crops. A few short years ago, the scenario of Roundup being administered to cropland via airplane was unthinkable.

Outcrossing is a migration of the transgene unintentionally to other species, subspecies, or strains. It may sound incredible that a piece of DNA could actually migrate, but documented cases have occurred. The industry claims that the risk of outcrossing has been exaggerated. However, recent discovery of a transgene in wild Mexican maize has renewed concerns. The scientific community is worried about the loss of biological diversity if all related species show up with the same GM genes in their DNA. A crop species without any genetic diversity might be vulnerable to catastrophic losses to a new pathogen. Opponents of this technology also point out that outcrossing of the herbicide-tolerant gene to the wild population could produce superweeds that would be extremely difficult to control.

In food crops, discovery of GM genes in non-GM crops really upsets people, even though there is no discernible risk to human health. The public is very suspicious and they want to know what they are eating. Transgenes have been discovered in crops that were sold as traditional in several cases. Substantial losses were sustained by the agricultural entities selling these crops. Concerns about the inadvertent or surreptitious dissemination of GM crops have resulted in stricter control of the GM crops, especially in Europe. However, in the United States, once these crops are approved, they are no longer subject to additional regulatory surveillance.

Insect Resistance

The main technology for producing *insect resistant* crops uses a natural protein produced by bacteria, called Bt, which stands for *Bacillus thuringiensis*, a soil bacterium whose spores contain a crystalline (Cry) protein. This protein breaks down in the gut of the insect and releases a toxin. The Bt toxin creates pores in the intestinal lining and kills the insect. Insecticides containing Bt, such as Dipel, Thuricide, VectoBac, have been available for many years and are thought to be safe for mammals, birds, and most nontarget insects. Bt-based insecticides are widely used by organic farmers because they are considered "natural." The link between Bt technology and organic farming creates a significant problem for the GM crop industry (see next).

The bacterial gene coding for the Bt protein, called the *Cy gene*, has been inserted into the DNA of corn, cotton, and potatoes so that the plant produces the toxin. In corn, it controls the European corn borer, the corn earworm, and the Southwestern corn borer. In cotton, Bt controls the tobacco budworm and cotton bollworm. These pests cost farmers millions each year in lost crops. The use of Bt has significantly increased yields (Carpenter, 2001). However, this technology is controversial because of the potential for resistance to Bt to develop in the target insect population. See the section "Bt Crops and Organic Farming," later in the chapter.

Another less controversial but less successful method for controlling insects has been to release insect biosaboteurs into the breeding pool. Genetically engineered bollworms that contain a lethal gene have been released into the wild. The hope is that they will breed with the wild population and all will die. A genetically engineered predator mite has been released in Florida to eat other mites that damage crops. These techniques are considered worth doing, but it is difficult to release enough insects to affect the natural population.

Pathogen Resistance

The engineering of resistance to viruses is another major thrust of GM crop research. Viruses take a sizeable fraction of our food every year. There are a number of examples where virus-resistant plants are in production, although admittedly, the use is a bit isolated. The tropical fruit papaya is susceptible to a number of pests and diseases. A transgenic variety of papaya, called UH Rainbow, is resistant to the papaya ringspot virus and is currently in production in Hawaii. Researchers at the University of Florida have produced grape vines carrying a gene derived from silkworms that protects against the fatal bacterial disease, Pierce's disease. Pierce's disease affects grapes and several other plants. Expression of a gene for an antibody against artichoke mottled crinkle virus in transgenic tobacco has successfully reduced infection in tobacco plants (Tavladoraki, P. et al., 1993).

Nutritional Enhancements

Some of the genetic engineering dollar has been invested in improving crop nutrition. These improvements have the potential to genuinely impact the health status of third-world populations. However, servicing an impoverished client is not a typical strategy for corporate growth. For the developments in nutritional value of cash crops to truly benefit those most in need, governmental or international organizations may have to provide incentives to companies owning the technology.

GOLDEN RICE

The poster child for *nutritionally enhanced* GM crops is golden rice. Golden rice is the general name for rice varieties that, unlike traditional varieties of rice, contain the precursor to vitamin A, beta-carotene. This technology was developed by Dr. Ingo Potrykus of the Swiss Federal Institute of Technology in Zurich and Dr. Peter Beyer of the University of Freiburg in Germany. The rice contains two daffodil genes and one bacterial gene that together carry out the four steps required for the production of beta-carotene in the rice. The resulting plants appear normal except that their grain is a golden yellow color.

The inventors hope that golden rice will alleviate vitamin A (retinol) deficiencies in the populous developing nations, many of whom use rice as their staple. Vitamin A deficiency is major cause of blindness in children from these regions. The World Health Organization estimates that between 100 and 140 million children are vitamin A–deficient. However, opponents of golden rice claim that the vitamin A content of golden rice is not high enough to alleviate the problem. Greenpeace claims that a daily intake of 300 grams (0.7 pounds) of golden rice would at best provide 8 percent of the vitamin A daily adult requirements. This means that a normal adult would have to eat 12 pounds of rice a day. Developers point out that the data on how much rice would be required to prevent blindness is incomplete, and that the rice is intended as a supplement, not necessarily to provide all required vitamin A. For this technology to work in different regions, the beta-carotene producing genes must be transferred to local varieties of rice. It will be a while before developing nations can benefit from this technology.

OTHER NUTRITIONALLY ENHANCED CROPS

Frustration with the quality of commercially available tomatoes has inspired home gardeners for generations. So, tomatoes have been a prime target for bioengineered enhancement. The bioengineered FlavrSavr tomato was the first commercially available crop nutritionally enhanced by GM technology. It was available for a long time before being discontinued. This tomato was engineered to ripen slowly. Because it remained on the vine longer, it exhibited improved flavor. However, it was based on a variety that was not competitive for other reasons, resulting in its ultimate failure. Under current development are tomato varieties with increased lycopene content (a component of vitamin A) and canola oil with enhanced vitamin E content.

Hardiness

Agriculture in many parts of the world is limited by the amount of arable land. And desertification (creeping desert) is diminishing the land that is available in the Middle East and elsewhere. There is considerable interest in producing plants that can grow under substandard conditions. Scientists at the University of California and the University of Toronto have developed a tomato plant that is able to tolerate high levels of salt and that holds the salt in its leaves, so the fruit will not taste salty (Zhang and Blumwald, 2001). This technology is being explored in other food crops. And perhaps one day the desert can be made to bloom without water!

On a less serious note, lawn grasses are being developed that are Roundup-resistant, grow more slowly, and need less water. This technology will save water in cities in drought-stricken areas, reduce the cost of maintaining recreational facilities such as golf courses, and free millions of homeowners on the weekends. Critics fear the production of a grass that nothing can kill and that could invade diverse types of habitat. We (humanity) have had a very bad experience with a strain of algae that was cultivated to use in fish tanks. This algae, *Caulerpa taxifolia*, is great for fish tanks because it is hardy and the fish won't eat it, but it is very bad once it is out in the ocean. Evidently, a fish tank owner or owners dumped their fish tank seaweed into the ocean, and now invasions of *Caulerpa taxifolia* have overtaken coastlands in the Mediterranean, Australia, and California (Withgott, 2002). A GM grass could create a land version of the killer seaweed.

Control of Seeds

Development of a GM plant is very expensive. One of the challenges for the industry producing transgenic crops is to continue to receive revenue from these crops after they have sold the seeds. Farmers have traditionally saved the seed from some of their crops for next year's planting. (Note that seed producers have never been

happy about this.) I am sure that you have heard the phrase "don't eat your seed corn." Future generations may not understand this phrase because there will be no option to horde "seed" for transgenic crops. Farmers will be required to buy new seed every year. This is not because the transgenic crops do not pass the transgene onto their offspring, for indeed they do. However, the companies producing the transgenic seed want to motivate the farmer to buy new seed every year. In many transgenic crops, the industry is developing methods to engineer self-destruction into the seed. This will prevent replanting as well as a phenomenon known as "volunteer" plants. These are plants that germinate from seeds that escaped during harvest of a previous season.

One recently patented method of destroying seed is the *technology protection system* (TPS) or *"terminator" technology*. "Terminator" is probably an unfortunate name because it makes a bad impression with the potential customers. This technology inserts a trait that kills developing plant embryos, usually by expression of a ribosome inhibitor gene. The technology is such that this trait is inactive for the first generation so that one generation of plants can be grown with their seeds. Farmers will need to buy new seed every generation. Another method under development is *trait-specific genetic use restriction technology* (T-GURT). T-GURT plants will require an annual application of a proprietary chemical to activate the transgenic traits, so farmers either pay for the new seeds or pay for the proprietary chemical.

Proponents of the terminator technology point out that outcrossing will be limited because wild plants pollinated by these transgenic plants will produce dead seeds. However, critics turn this argument around by pointing out that the transgenic gene could kill neighboring seed crops or wild seeds. The industry is attempting to develop terminator crops whose pollen is free of the terminator gene. Some verbiage in the terminator patent raised some hackles, like the potential to use diphtheria or other toxins instead of the ribosome inhibitor gene, and the potential to use tetracycline in the process, increasing the presence of this antibiotic in the environment. Other concerns involve the additional expense to poor farmers in developing nations. And human nature is such that relinquishing control of the seed stock for global agriculture to a few mega-companies, well, makes people nervous.

The T-GURT technology may be an easier sell. With this technology, the seed doesn't die; it just doesn't express the transgene unless it receives special treatment. This gene would not be expressed in the wild and poor farmers could still use their seed, only without the transgene. Of course, poor farmers could always use seed without the transgene. The potential for migration into the wild, expressed or not, still causes concern. The chemical sprayed on crops to activate the transgene could have undesirable environmental effects.

Other Controversies Involving GM Crops

In addition to the concerns noted above, the implementation of GM technology in agriculture has raised concerns about impacts on organic farmers, wildlife, human health, third-world countries, and even the climate. Several examples that follow GM technology has given rise to unexpected controversy, for various reasons.

BT CROPS AND BUTTERFLIES

A highly emotional debate was initiated when researchers published results from laboratory experiments that demonstrated that pollen from Bt transgenic corn kills larvae of the monarch butterfly (Bouchie, 2000). As GM crops take over globally, we appeared to face a monarch butterfly–less world. However, a recent EPA study and others concluded that the probability of adverse effects of Bt corn pollen on monarch larvae was small (Dale et al., 2002, and Mendelsohn et al., 2003). However, the same report raised a concern about other members of the phylum, Lepidoptera, which includes butterflies and moths. It ultimately concluded, however, that primarily because of habitat differences, the risk to these creatures was small. Interestingly, whatever the emotional appeal of the fate of the butterfly, this issue had little effect on the implementation of Bt technology.

BT CROPS AND ORGANIC FARMING

Organic farmers believe that they have a lot to lose with the emergence of GM crops. At a minimum, the existence of this technology certainly complicates the definition of "natural." Organic farmers are especially concerned that the wider use of Bt in insect control will result in the evolution of Bt-resistant insects. If this happens, the organic farmers, who steadfastly resist GM technology, will not be able to use their "natural" insecticide. You may wonder why the issue of resistance has not been raised before, when organic farmers were liberally using the Bt toxin on their crops. The reason for this is that when the use of this technology becomes essentially global, the evolutionary pressure for development of resistance will be much higher because nonresistant insect strains survive anywhere. In response to these concerns, farmers growing the GM crops are required to reserve 20 percent of their acreage for "refuge," or traditional crops. The thinking is that the insects feeding on the traditional crops will retain the Bt-susceptible genome and will prevent the development of resistance. This strategy will work only if farmers worldwide comply with the target acreage requirement, a requirement that will certainly cost them in productivity of their acreage. Many people are skeptical about

a crucial control that will cost the farmer money. The other issue is a technical one. The "refuge" crop strategy assumes that the susceptibility trait will be the dominant one, so that if a susceptible insect breeds with a resistant insect, the offspring will be susceptible.

The EPA is implementing an insect-resistance management (IRM) program to alleviate these concerns. In addition to the "refuge" requirements just described, the EPA is monitoring insect resistance and has required remedial action plans from companies that outline measures they will take should insect resistance be detected (Mendelsohn et al., 2003). Whatever the effectiveness of these measures on this continent, the fact that this is a global issue gives one pause for thought. The technology will be disseminated into areas where government regulations are likely to have little effect.

CY GENE AND OUTCROSSING

Another concern with the Cy gene is outcrossing. Outcrossing greatly increases the chance of evolving insect resistance. The main factor that affects how likely a gene is to move into the wild is the colocation of wild plants that are closely related to the engineered crop. The EPA has studied the potential for expression of the Bt endotoxin by relatives of corn, cotton, and potatoes in the United States. It concluded that the products currently registered are at low risk for transfer to wild plants, either because their pollen is incompatible with the wild plants or because they don't share the same habitat. However, the EPA did recognize some potential for cotton genome migration in specific areas. In these areas, the sale or distribution of Bt cotton has been restricted (Mendelsohn, 2003). Other parts of the world may not enjoy a physical barrier between the cultivated and the wild species. These places tend to be much less regulated than the United States.

STARLINK CORN

In October 2000, Kraft Foods had to withdraw taco shells from supermarkets worldwide because genetically engineered corn approved for animal use only was found in these shells. Later, news reports revealed that the same corn (StarLink) had been found in other food products. Incidentally, the modification of the StarLink corn was a transgene to produce a Bt protein. The media tended to describe this corn as "unfit for human consumption," which was not true but not necessarily untrue: it had not been approved for human use. The parent company, Aventis, initially petitioned the U.S. government to temporarily approve the corn for human consumption, thus correcting the problem overnight. However, the FDA, sensitive to its potential loss of credibility, was unsympathetic. Aventis bought back the corn at a price of millions of dollars. As a consequence, the EPA will no longer approve GM products for animal consumption only.

The drama spread internationally as food shipments from the U.S. were rejected upon fear that they were contaminated by the GM corn. For example, a consignment of 1,000 tons of corn-soy blend, commissioned by the Catholic Relief Services, was sent home by the Indian government because the U.S. government refused to certify that it was free of StarLink (Jayaraman, 2003).

IMPACT ON THIRD-WORLD ECONOMIES

The globe has already seen an example of the negative effect GM crops may have on the developing world, especially in areas vested in just a few agricultural products. Malaysia is one of the world's largest suppliers of palm oil obtained from the kernel of the palm oil seed. One of the most valuable of the palm oil products is lauric acid. The Philippines also exports a large amount of lauric acid, derived from coconut oil. Calgene, a U.S. plant biotechnology company, has developed a genetically engineered rapeseed plant that produces an oil composed of a large amount of lauric acid. This method of producing lauric acid will be much cheaper than using palm oil. Since rapeseed is produced all over the world, the economy of Malaysia is at great risk if the market turns to the rapeseed-produced lauric acid. Malaysia, as one of the world's largest producers of natural rubber from trees, is also facing the potential of much more productive GM rubber trees, trees that are likely to throw some of the population out of work (Bourgaize et al., 2000).

THE INTERESTING CASE OF ICE MINUS

Pseudomonas syringae is a very unpopular bacterium with produce farmers because it secretes a protein that causes ice crystals to form on their produce. It is therefore a major factor in frost damage to crops. A genetically modified form of *Pseudomonas syringae* was developed in the mid 1980s that lacks this protein, hence the name "ice minus." The proposal was to spray the ice minus bacillus on crops at risk to frost damage.

Field trials were conducted in California and the U.K. to establish the safety and effectiveness of ice minus technology, but it is still not used because litigation over its use is tied up in courts all over the world. The concern is that use of ice minus may alter the climate. Scenarios of bacteria escaping into clouds and preventing precipitation have been disconcerting enough that the amount of evidence needed to prove the negative (that it won't happen) keeps growing. This potential was sufficiently famous to evoke comment from the Vice President of the United States.

What is most interesting is that the opposite use of this type of technology—i.e., use of nucleating proteins to create "artificial" snow—has not raised a peep of opposition. Perhaps the potential impact on the climate—creating precipitation—would not be considered a hazard.

Future GM Crops

Essentially, only our imagination limits the potential for new GM crops. We are going to be very lucky consumers indeed as we are provided with food that tastes wonderful, lasts longer, and is easier to cook. Among the technologies likely to have the most impact is the effort to engineer new food crops that can take in nitrogen directly from the air rather than having to rely on fertilizers.

Trees are being modified to provide resistance to insects, tolerance to herbicides, and a higher yield of the commercial product. For example, reducing the lignin content of a tree can make it easier to recover wood pulp. And we are likely to see such commercially appealing products as nicotine-free tobacco.

The use of plant grafts and plant shoots is likely to become more widespread for propagation of GM crops. This is a very traditional method of cloning, and of course can be used to clone transgenic crops. Experiments are underway to produce vanilla, orange, and lemon vesicles from GM tissue culture.

Bioengineering of Livestock

There are currently only two genetically engineered food animals designed for human consumption (Dove, 2005). One is the transgenic salmon developed by Aqua Bounty Technologies. The growth hormone gene from Chinook salmon has been spliced to a promoter; the promoter causes the gene to be transcribed. Unlike the natural fish, the transgenic fish express the growth protein year-round, significantly increasing their growth rates. This trait is being applied to other farm fish, including trout, tilapia, and turbot. Environmentalists have expressed extreme concern that such bioengineered fish could readily push out natural fish population, pointing out that farm-raised fish routinely escape into the wild.

The second transgenic farm animal is the Enviropig, developed by researchers at the University of Guelph in Ontario. This animal has been developed with a gene for phytase, expressed through the salivary gland. Phytase breaks down phosphates in the pigs' food, reducing phosphorous excretion in the animals' waste. Widespread use of these animals could significantly reduce the environmental impact of pig farms.

There are many promising research efforts underway to bioengineer animal food products. A few engineering efforts have focused on reducing susceptibility to disease among farm animals, including studies in chickens, sheep, and cows (Wall et al., 2005). The company ViaGen has identified shrimp that carry disease-resistance genes and have selected animals carrying the gene. Development of disease resistant strains of shrimp has a potentially huge impact on the Pacific Rim, where shrimp farming is big business. Researchers at the University of Minnesota are investigating genes in turkeys that cause the development of larger breasts and are associated with resistance to Salmonella.

Just as we humans are interested in our ancestry, livestock producers are interested in biotechnology that helps identify an animal's lineage. For example, in the beef industry, Angus lineage is valuable but difficult to certify. ViaGen recently capitalized on this with a genetic test that generates a percentage-Angus score for any bovine blood or tissue sample.

Federal and private funding for bioengineered farm animals has lagged behind other agricultural initiatives. However, several companies are developing animals that grow faster, including pigs, sheep, and brooding turkey hens. Other thrusts are in the areas of genetic profiling and cloning. However, cloning has a low production efficiency (below 2 percent) and a high incidence of anatomical defects in the cloned offspring. One technology that has received public attention is the use of Bovine Growth Hormone (BGH) to increase both growth rate and milk production in cattle. This practice is reported to increase the incidence of mastitis and consequently to increase the use of antibodies, raising concerns about the milk supply. Also, there are indications that increasing BGH levels shorten a cow's life and make individual animals more susceptible to injury.

We can expect to see acceleration of bioengineering of farm animals with the completion of the sequencing of the chicken and the cattle genome. This knowledge will form the basis for rapid changes in livestock animals, how they are produced, how big they get, and how their products taste. Table 11-1 shows the benefits and risks of biotechnology applied to agriculture.

Feeding the World

The biotechnology industry has engaged in a largely successful public relations campaign that emphasizes the pressures that the human population is putting on the food supply, and has promoted the notion that biotechnology holds promise for feeding the teeming masses, at least for now. However, critics point out the major challenge in feeding the people of developing nations traditionally has been food distribution rather than food production. In addition, as in the case of

Table 11-1 Summary of the benefits and risks of agricultural applications of biotechnology

Technology	Positives	Risks
Transgenes—general	Potential to enhance all forms of human agriculture.	Decrease in biological diversity with potentially greater impact of new pathogens; migration into the human food supply without the knowledge of the consumer; technology may not thrive in new locations.
Herbicide resistance	Increased agricultural productivity; decreased use of nonbiodegradable herbicides.	Outcrossing, resulting in the development of super weeds, resistant to herbicides.
Insect resistance	Increased agricultural productivity; decreased use of nonbiodegradable insecticides.	Outcrossing, resulting in the development of Bt-resistant insects. Bt would no longer be an option for organic farmers. Adverse affect on butterfly populations.
Nutritional enhancements	Positive impact on nutritional status for people in developing nations.	Might actually have little effect but could serve as an advertising hype for commercial entities.
Hardiness	Allows agriculture on marginal land.	Development of super-hardy plants that could become nuisance plants, such as invasive grass.
Frost resistance	Increased agricultural productivity.	Might result in climate change.
New product sources, such as lauric acid from rapeseed plants	Decreased cost of production of some plant products.	Wreak havoc in economics dependent on traditional production methods.
Bioengineered farm animals	Increased productivity of animal-based food products.	In the case of bioengineered fish, farmed fish might escape and overwhelm native populations.

Malaysia, GM crops may introduce economic factors that increase hardships in these areas. We will need very, very wise leaders who communicate well internationally for biotechnology to reach its potential in alleviating human suffering in the area of agriculture.

Summary

The use of genetically modified (GM) crops is widespread in the U.S. and internationally. Thus far, the major crops are soy, corn, and cotton. However, many other GM crops are either being implemented or are under development. The major modifications are in the area of herbicide tolerance, insect resistance, and pathogen resistance. Other upcoming modifications will improve the tastiness, nutritional value, and general hardiness of plants.

Herbicide resistance is primarily resistance to the broad-spectrum herbicides, Roundup and Liberty. Insect resistance is usually incurred by inducing the plant to produce the insecticide protein Bt, which is produced naturally by a bacteria. These technologies have raised concerns that the new genes will migrate into the wild population or into other food crops, creating insects resistant to Bt and super weeds. Organic farmers are especially concerned about the Bt technology because they use the Bt toxin on their crops as a "natural" insecticide. The global exposure of the insect population to this toxin creates huge evolutionary pressures for development of resistance. In this country, the EPA has taken measures to monitor the insect population and has required farmers growing GM crops to devote a fair portion of their acreage to "refuge" or natural crops. Other concerns are the effect on wildlife, like butterflies. EPA studies to date indicate that these concerns are not well-founded.

There have been some spectacular failures in agricultural biotechnology. By and large, these have been failures in public perception. Among the most famous is the StarLink episode, which cost the parent company $100 million to recall GM corn that had been approved only for animals but was found in the human food supply. Another was the "ice minus" bacteria, which raised an issue regarding potential climate change and whose litigation is still before world courts.

The developing world has the most to gain and the most to lose from GM crops. Nations currently unable to feed their burgeoning populations could vastly improve the quantity and quality of their food supply through biotechnology. However, GM crops also have the potential to replace traditional agricultural crops. And in a global economy, the replacement may not occur locally.

Quiz

1. The Bt technology could harm Monarch butterflies because:
   (a) the Bt toxin kills the plants the butterflies feed upon.
   (b) the Bt toxin could migrate to the wild plant population.
   (c) the monarch larvae feed upon plants that include GM plants with the Bt toxin.
   (d) all are correct.
   (e) b and c are correct.
2. Ice minus could prevent frost damage to crops by:
   (a) coating the plants with a frost-resistant coat.
   (b) generating heat from the metabolism of the bacteria.
   (c) eliminating a protein that serves as the nucleus for ice crystal formation.
   (d) changing the climate and inducing global warming.
3. Use of GM crops globally:
   (a) has been limited due to public opposition.
   (b) is especially unpopular in the Far East.
   (c) has been widely embraced for a few crops.
   (d) has the potential to disrupt incomes based on traditional agricultural products.
   (e) c and d are correct.
4. The planting of "refuge" acreage:
   (a) is part of traditional organic farming.
   (b) is required by the EPA.
   (c) will prevent soil erosion.
   (d) will help prevent development of insect resistance to insecticides in GM crops.
   (e) a, b, and d are correct.
   (f) b and d are correct.

5. Outcrossing is:
   (a) a potential benefit of GM agriculture.
   (b) can be prevented by the use of T-GURT technology.
   (c) can be prevented by "refuge" acreage.
   (d) is not possible with current GM crops.
6. Terminator technology:
   (a) is motivated by a desire to compensate seed-developing companies for the seed.
   (b) would prevent collection for replanting the next season.
   (c) kills the seed at crucial points in embryo development.
   (d) raises concerns about killing natural populations due to outcrossing.
   (e) all are correct.
7. Transgenic livestock:
   (a) includes pigs that have less environmental impact than their natural cousins.
   (b) currently includes only two species that are marketed commercially.
   (c) includes many species under development for improved productivity and disease resistance.
   (d) includes species grown in fish farms.
   (e) all are correct.
   (f) c and d are correct.
8. Transgenic fish:
   (a) are routinely released to improve the native stock.
   (b) have not been approved for commercial development.
   (c) have been engineered to grow larger.
   (d) are not likely to escape fish farms because of an engineered behavioral trait.

9. Cloning in animals:
   (a) is routinely used to reproduce a very successful livestock line.
   (b) suffers from a very low productivity.
   (c) experiences a high rate of birth defects.
   (d) can theoretically be used to propagate a transgenic herd.
   (e) all are correct.
   (f) b and c are correct.
10. Organic farmers:
    (a) are using transgenic crops that express the natural Bt toxin.
    (b) have special-interest concerns about the development of Bt-resistant insects.
    (c) plant "refuge" acreage among their crops.
    (d) all are correct.

References

Bouchie, A. 2000. Bt corn kills monarch? *Nature Biotechnology* 18:1025.

Bourgaize, D., T. R. Jewell, and R. G. Buiser. 2000. *Biotechnology: demystifying the concepts*. San Francisco: Benjamin Cummings.

Carpenter, J. E. 2001. Case studies in benefits and risks of agricultural biotechnology: roundup ready soybeans and Bt field corn. National Center for Food and Agricultural Policy, Washington D.C. http://www.ncfap.org./pup/biotech/benefitsandrisks.pdf.

Dale, P. J., B. Clarke, and E. M. Fontes. 2002. Potential for the environmental impact of transgenic crops. *Nature Biotechnology* 20:567–574.

Dove, A. W. 2005. Clone on the range: What animal biotech is bringing to the table. *Nature Biotechnology* 25 (3): 283.

Jayaraman, K. S. 2003. US food aid still under GM cloud. *Nature Biotechnology* 21:346–347.

Mendelsohn, M., J. Kough, Z. Vaituzis, and K. Matthews. 2003. Are Bt crops safe? *Nature Biotechnology* 21:1003–1009.

Taverne, D. 2005. The new fundamentalism. *Nature Biotechnology* 23:415–416.

Tavladoraki, P., E. Benvenuto, S. Trinca, D. De Martinis, A. Cattaneo, and P. Galeffiet. 1993. Transgenic plants expressing a functional single-chain Fv antibody are specifically protected from virus attack. *Nature* 366:469–472.

Wall, R. J., A. M. Powell, M. J. Paape, D. E. Kerr, D. D. Bannerman, V. G. Pursel, K. D. Wells, N. Talbot, and H. W. Hawk. 2005. Genetically enhanced cows resist intramammary *Staphylococcus aureus* infection. *Nature Biotechnology* 23:283–285.

Withgott, Jay. 2002. Invasive species: California tries to rub out the monster of the lagoon. *Science* 295 (5563): 2201–2202.

Zhang, Hong-Xia, and E. Blumwald. 2001. Transgenic salt-tolerant tomato plants accumulate salt in foliage but not in fruit. *Nature Biotechnology* 19:765–768.

CHAPTER 12

Industrial and Environmental Applications

Mankind has been harvesting the products of cellular manufacturing processes since the dawn of civilization—since humans first realized that grapes infested with yeast and allowed to ferment produced wine. We are becoming masters at analyzing the manufacturing genius of the cell and modifying cellular production methods to serve industrial needs of our own. We have taught cells to produce industrial quantities of organic chemicals—chemicals that you would swear would be toxic to living organisms. Microbes can produce most carbon-based structures that we need. Currently, petroleum is fairly inexpensive, and most of the organic chemicals that we use are derived from oil. Eventually, bioproduction will compete with petrochemical methods as the most economically feasible way to obtain the organic chemicals that we depend upon.

The cell is an exquisite set of manufacturing, destruction, and demolition processes—very much like a manufacturing plant. The purposes of the cellular manufacturing process are to extract energy from the environment and use this energy to sustain cellular existence, to ward off invaders, and to reproduce. These processes are so complex and intricately interrelated that the comprehensive system definition of the cell is currently beyond human capabilities. However, we have glimpses of discrete parts of the cell's processes, as if we were spying on the manufacturing plant, using a microscope to hone in on small sections of the plant. And we are stealing the secrets that we learn with our microscope and attempting to rebuild the plant from what we know.

We want to harness the secrets that the cell uses to produce the myriad of natural cellular products in an unnatural world and to produce them quickly and with few waste products. The secret relies on cellular enzymes. Enzymes typically accelerate biochemical reactions by 10^{10} (Salsh, 2001), a feat that you could not expect from your average organic chemist.

It would be wise to review the nature of enzymes. Remember that *enzymes* are a subset of proteins—proteins that act as *catalysts*. Catalysts are molecular chemists that speed up or alter chemical reactions. Enzymes include catalysts that can break down other proteins and catalysts that can serve as switches, moving chemical reactions in different directions. These are extremely efficient orchestrators of numerous biochemical reactions, many of which produce products of interest to humans. Thousands of enzymes exist naturally.

Use of Bioreagents in Industry

Chemicals that are captured from living cells and used as chemical reagents in production processes, either within the cell or outside of the cell, are known as *bioreagents*. Current widespread use of bioreagents can be found in the textile and food industry, and in pharmaceuticals.

A number of cellular enzymes are found in industrial chemical assembly lines and participate in the generation of thousands of products. Among the numerous examples of processes involving bioreagents in the pharmaceutical industry is the biotransformation of natural plant steroids to produce contraceptives and other steroid hormone derivatives. The production of most of the common antibiotics involves the use of cellular enzymes. In the food industry, the use of enzymatic catalysis includes such familiar examples as the production of high-fructose corn syrup. The low-calorie sweetener aspartame is produced on a kiloton scale by Holland Sweetener Company using a proteolytic enzyme, thermolysin. Industrial applications include the production of phenolic resins, acrylamide, and many modern insecticides. The "natural enzymes" that are included in detergents come from cellular enzymes.

Finding Suitable Bioreagents

It is not easy to sort out the bewildering array of cellular chemicals and find a particular enzyme. You have several options:

- Starting with a cell actively performing the function of interest, isolate enzymes at random and examine the amino acid sequence of each one. Compare to structure of enzymes with known function to identify similarities. You may stumble upon the enzyme with the function you need.
- Isolate mRNA from cells actively producing the desired product and look at the enzymes that the mRNA translates into. Concentrate on enzymes whose presence is dependent upon cell activity. Again, compare to enzymes with known functions to identify similarities.
- Using a population of inactive cells, experiment with the process by adding enzymes at random and selecting for cell populations that begin to display the correct activity. You can conclude that the enzyme added to the newly active cell population is critical.

Sounds like a lot of work? These approaches are very labor-intensive, and the discovery of new enzymes has tended to be serendipitous. Historically, the bioenzymes used in industry were fairly limited and produced by a few old-faithful, well-known organisms, such as E. coli.

One factor enabling a research explosion in this area has been the development of huge databases on protein structures and sequences. These structures have been analyzed to identify commonalities among proteins with similar functions. Hundreds of thousands of natural enzymes occur in a limited number of archetypes or *superfamilies* throughout the biological kingdom. Superfamilies have similar structure and similar function and are found across distantly related species. A new enzyme can be compared structurally to others whose function is known, and the probable niche this enzyme fills can be extrapolated. Identifying the family for your enzyme or even matching with enzymes with very similar amino acid sequences doesn't necessarily solve your problem, however. Remember that the function of proteins depends on their three-dimensional structure and that the change of just a few amino acids can foul up an otherwise perfectly functional protein. Nonetheless, the use of computerized computational methods has accelerated biochemical screening within the last decade and has resulted in the identification of a burgeoning population of potentially useful bioenzymes, accounting for part of the patent frenzy. You can now add catalytic RNA and even DNA catalysts to the list of cellular chemicals of interest to you. Figure 12-1 summarizes the process for finding cellular enzymes.

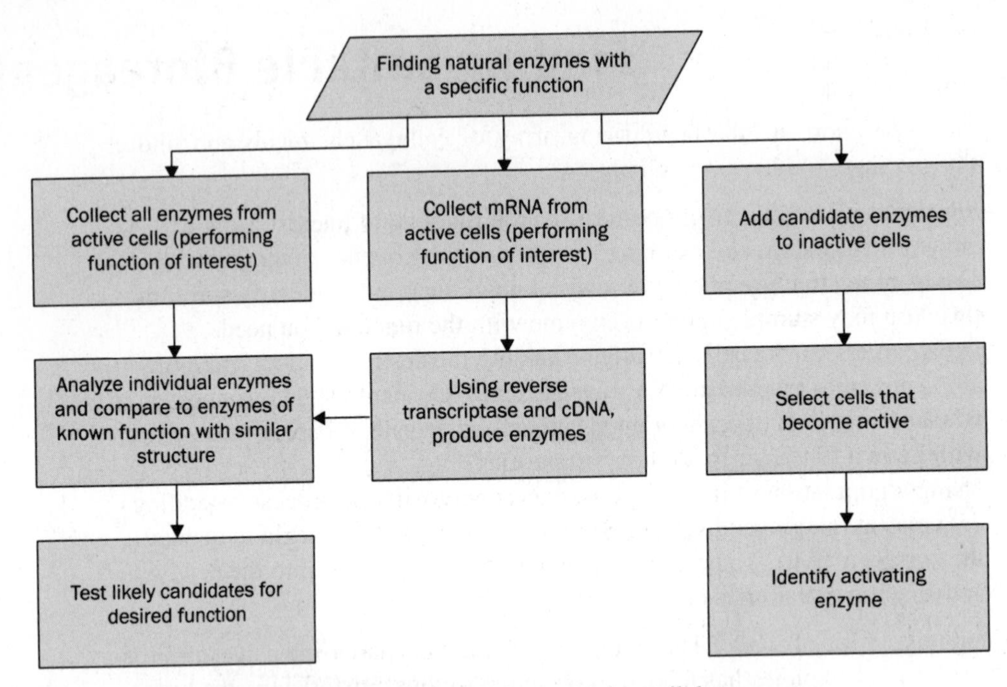

Figure 12-1 Finding and characterizing cellular enzymes

Modifying Enzymes and Molecule Sex

Natural enzymes are incredibly efficient, but the cell has a different set of priorities than we do. We want to crank out large (industrial) quantities of a given product. Cells want to closely regulate their environment and conserve energy by producing only what they need when they need it. Many cellular enzymes are fragile, heat-labile and short-lived. Furthermore, they tend to be "turned off" by the accumulation of the products of the reaction they enable. Sometimes the enzymes we need are not available in nature, so we have to modify other enzymes to be useful to us. Figure 12-2 summarizes the methods of producing new enzymes for industrial use.

Happily, increased understanding of enzyme function has lead to recombinant DNA methods to tailor these functions to human needs. We can re-engineer what nature has given to us. For example, we can look at enzymes from organisms that function at temperature or pH extremes and investigate ways to incorporate the characteristics of these hardy enzymes into our production process enzymes. We are especially interested in increasing the specificity of a given enzyme—to limit it, for example, to a specific stereo configuration of its target. On the flip side, we would like to engineer enzymes with multiple active sites and multi-enzyme systems that can catalyze multistep metabolic processes.

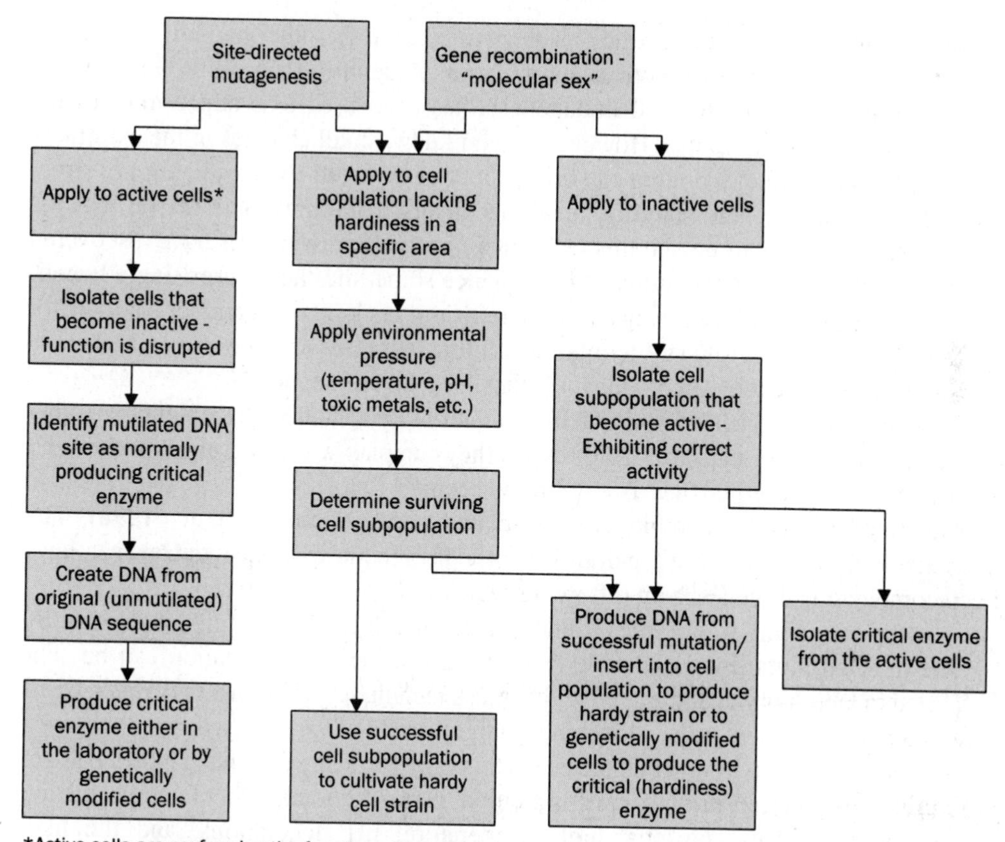

Figure 12-2 Derivation of new bioreagents and bio-factories

One might expect that recombinant DNA technology has increased the availability of natural enzymes for industrial use. Development and innovative use of new enzymes using recombinant DNA technology are still largely found within the academic community. However, given the recent explosion in knowledge about cellular processes and the accompanying explosion in patent applications, it is obvious that the use of bioreagents in manufacturing processes is about to erupt with a volcanic force. The statement of Schmid et al. (2001), "The use of biocatalysis for industrial synthetic chemistry is on the verge of significant growth," may be somewhat of an understatement.

One method of tweaking the composition of a given molecule is called *site-directed mutagenesis.* This is a method of inducing a random mutation at a specific site in the DOE code. In one application of this technique, the researcher observes the effect of the mutation on cell function and can work backward to determine the source of the effect. If a particular cell function is disrupted, the researcher can conclude that the protein whose code was disrupted is responsible. On the other hand, if an inactive cell is induced to become active, the mutation evidently results in production of a critical enzyme. Site directed mutagenesis has been used to produce most of the engineered enzymes in use. However, as you know, small changes in the extremely complex structure of a protein can bring about unexpected changes in tertiary structure, functionality, and stability, resulting in lots of failures. The reason that the industry continues to pursue this technique is because researchers have discovered that small changes in the amino acid sequence sometimes have surprisingly beneficial effects, such as fine-tuning a desired reaction. It's kind of a gambler's fallacy for the researcher. That is to say, it only takes a few successes to motivate researchers to put tremendous energy in random searches for a given function.

In addition to artificially inducing mutations, researchers have found another method to increase the variety of proteins they can play with, creating a researcher's sandbox. This method has been fondly deemed "molecular sex," and refers to laboratory-induced recombination of related genes (Crameri et al., 1998). The methodology results in a random increase in genetic diversity and even allows recombination of genes from different species.

After the variety of enzymes found in a given cell has been augmented either by site-directed mutagenesis or gene recombination, the cell population can be subjected to laboratory evolution. Note that these techniques result in a very diverse cell population. The new cells, producing a large variety of randomly derived proteins, are subjected to environmental pressures that only the lucky few that happened to get a useful enzyme can survive. By this method, researchers can select for cells that are tolerant to metallic poisons, high temperature, pH fluctuations, and the like. Researchers can select for cells that don't exhibit negative feedback and produce large quantities of the product in question. The increased activity and stability may make a marginal process economically feasible.

In addition to engineering for better activity, improved substrate range, and stability in the enzymes, the host cell can also be engineered to eliminate side reactions and to export its products. A huge technical challenge in producing bioengineered products involves harvesting the product from the cells. Therefore, getting the cell to release the desired product is of great significance.

You can also actively engineer new proteins by random combination of amino acids, coming up with amino acid sequences the world has never seen before. For this to be useful, you would need to have an idea of what you need structurally. You can base your educated guess on the huge amount of information regarding proteins and their functions. However, analyzing this information is impossible without the use of modern computational methods.

An especially challenging technology is *pathway engineering*, whereby the combinations of genes that create a complex product are determined. Frequently, complex biochemical pathways involve more than one type of cell. Genes from these

cells can theoretically be compiled and introduced into a single host. The development of golden rice is an example of successful pathway engineering. Genes from several unrelated species that code for different intermediates in the production of beta-carotene have been combined in a single species to produce a rice containing beta-carotene.

Inside or Outside of the Cell

You are an organic chemist designing your biological manufacturing process. You can isolate the enzymes of interest or make them synthetically, and make your product in a reaction vesicle of some sort. Or, you may elect to use the cells as your manufacturing plant. Some of the things you need to consider are listed here:

- Are all the processes needed confined within a single cell? If not, you need to carry out your process outside of the cells or use pathway engineering to put all the genetic codes into one cell.
- Will the reaction be carried out under conditions survivable by living cells? If not, the reaction should be conducted outside of the cell in the laboratory, or a hardier cell population must be derived.
- Are there cofactors that must be regenerated during the manufacturing process? That is, does the cell need to modify some of the reagents during the process? If so, you will need the living cell to perform this operation.

Are you sure you understand all of the factors involved in successfully making the product? If you are dealing with living cells, you need to consider all of the factors that hamper mother nature in developing a viable population, including temperature, pH, limits on cell density, the need to eliminate waste products, etc. If you fail to supply the cell with the things it needs to live, in addition to the things it needs to produce the product, your project will fail. For example, you might insert a gene into a cell that produces a product but also creates substantial waste. If the waste adversely affects the viability of your cell population, you are not going to be able to produce your product.

Bioreactors

The vessels where biosynthetic operations occur are called *bioreactors*. A bioreactor may contain either a large quantity of living cells or a cell-free mixture of reagents and enzymes. Most biocatalytic processes operate in aqueous environments and are usually operated in *batch mode,* which is a one-time operation. The cells or reagents are induced to produce the product. The process is terminated and the product collected. The alternative is a throughput operation that works continuously with periodic harvesting of the product. In a typical batch fermentation process, the desired cell is placed in a medium consisting of water, nutrient chemicals, a source of nitrogen, and dissolved oxygen. Cells are allowed to grow to a desired culture density. The liquid medium must contain all the chemicals required by the cells and also provide controlled environmental conditions. There are different strategies to circulate the life-sustaining medium around the cells and to eliminate CO_2 and other waste products. Typical methods are (1) using continuous stirring or (2) suspension of the microbe population in the circulating medium. Remember that dissolved oxygen must be continuously delivered to the cells. The cells produce the desired product(s) from precursor chemicals introduced into the nutrient mixture. The desired product(s) are then either excreted from the cells and purified from the liquid medium or chemically extracted from the cells.

There are a number of disadvantages to batch processes used on a commercial scale. They cost energy because of the requirement to maintain a large volume of fluid at the correct temperature. In addition, because cells are sensitive to such parameters as nutrient concentration and pH, the process must be continuously monitored and adjusted to maintain optimal conditions. The process is vulnerable to contamination and is limited by the accumulation of waste products. A significant problem is the fragility of whole cells. The cells are disrupted by high oxygen pressure and vigorous mixing methods, so you have to be gentle. As the cell density increases, delivering adequate dissolved oxygen (or any other gas) and nutrients to the cells becomes more difficult. Furthermore, batch processes often result in low product concentration.

The high costs associated with batch processes have resulted in the development of strategies to conduct these reactions in a much smaller volume that is easier to heat, circulate with nutrients, and sterilize. Rather than suspending cells in media or stirring the whole batch, cells are immobilized and media is circulated around them. Immobilization strategies include cultivating cells in polyacrylamide or agar gel matrices. Also, cells and biocatalysts will adhere to the external surface of spherical beads. These beads can either be fixed at the bottom of the reaction vessel or floated in the media. Once you are free of the big aqueous mixing vat, you can design much more efficient processes.

A new area of enzyme research has revealed that enzyme function can be improved by using them in organic solvents (Klibanov, 2001). Classic wisdom says that enzymes are denatured in solvents—everybody knows that. However, if the solution is entirely free of water, enzyme structures survive solvent immersion. Furthermore, their selectivity may actually be improved or may be completely altered. The use of enzymes in solvents allows them to act on the many organic compounds that are insoluble in water. Advances in bioengineering have lead to the

development of cells that are tolerant of solvents and can be used to produce enzymes in the nonaqueous environment.

Environmental Applications—Hazardous Waste

The removal of chemical contamination from water and soil has cost society significant amounts of money and has met with relatively limited success. "Superfund" sites continue to exist decades after their identification, and much of the money spent at these sites has resulted primarily in the generation of paper, reports detailing the extent of contamination, reports on the risk posed by the site, and reports on the risks posed by remediation. Options for removal of chemical contaminants are (1) the physical removal of the contaminated stuff, usually just from one place to another, or (2) the transfer of the contaminants from one medium, such as soil or water, to another, usually air. Many groundwater "pump-and-treat" processes strip the contaminants from the water and release them into the air. Even if the contaminated water is passed through granular activated carbon (GAC), the toxicity or absolute amount of contaminants is not reduced but simply moved or concentrated. High temperature incineration breaks down the contaminant but is very expensive and can produce toxic air emissions.

Bioremediation offers another option: the use of microbes at the site or at the hazardous waste treatment plant or landfill to convert chemical compounds into innocuous or less harmful chemical compounds. Bioremediation techniques enjoy very kind treatment by the popular press, and are promoted by the environmental community for good reason. However, there are a number of technical obstacles to overcome; and the hazardous waste industry continues to take the path of least resistance: dig it up and then bury it in a landfill or burn it. A number of recent developments promise to make bioremediation a more attractive option.

The most successful bioremediation technologies are those that utilize microorganisms that are naturally present in the contaminated media. Many naturally occurring bacteria and fungi are capable of degrading chemical contaminants, including microbes that can degrade pesticides and polychlorinated biphenyls (PCBs) and those that can detoxify metals. The other option is to introduce cultivated or GM microbes. However, the use of native species currently is the more attractive option.

Clearly, using natural process to chip away at hazardous chemicals will not progress in a timescale that we can accept. Consequently, we have thought of ways to speed the process up. *Biostimulation* refers to methods to jump-start the microbes that are already in place by fertilizing with nitrogenous and other nutritious

compounds. For example, methane has been injected into contaminated aquifers to act as a source of nutrients for the resident bacteria.

Other techniques have been devised to increase the *bioavailability* of insoluble contaminants, such as PCBs. Bioavailability is an index of how accessible a given chemical is to the plants and microbes in an area. High bioavailability is in general a bad thing in contaminants because they tend to move through the food chain. However, if you are attempting to improve the access of microbes to chemicals, you need to devise methods to increase chemical solubility and to keep the chemicals from absorbing onto soil particles. Actions that are helpful include application of heat, application of surfactants, moving moist air through the compost to keep the reaction media moist, and use of chemical reducing agents. Recent patents have involved the use of fibrous, cellulose material; legume-related plants or microbes; and fiber-degrading plants as part of the soil compost. The contaminants adhere to the fibers and exhibit increased bioavailability. One variable that is particularly important is the maintenance of anaerobic (oxygen-free) conditions either in situ (as you find it) or in the treatment process. Anaerobic conditions seem to favor detoxification of recalcitrant organic and inorganic contaminants in the environment.

Bioremediation is even an option with metal contamination. Metal-eating bacteria occur naturally in soil. The metal contaminants may include heavy metals such as arsenic, antimony, beryllium, cadmium, chromium, copper, lead, mercury, iron, manganese, magnesium, radium, nickel, selenium, silver, thallium, and zinc. Examples of applications include the remediation of heavy metals; the remediation of arsenic-impacted surface water, groundwater, and/or soil; the remediation of acid mine drainage; and the treatment of spent metal-plating solutions. Obviously, the elemental metal doesn't go away, but it is transformed to a less toxic form. The "doesn't-go-away" part tends to be unacceptable to the nearby inhabitants, unfortunately.

Among the most famous of the metal-eating bacteria are anaerobic bacteria that reduce hexavalent chromium to trivalent chromium. Hexavalent chromium causes cancer in humans where trivalent does not. Usually, chromium-contaminated soil is dug up and treated in a bioreactor so that conditions for microbial action can be optimized. Several recent technologies advise leaving the chromium-contaminated soil in place and adding reducing agents and certain nutrients that are enjoyed by the chromium-reducing bacteria. It always helps to keep the soil moist.

However, bioremediation at the site has presented some difficulties, especially in treating contaminated aquifers. The conditions in an aquifer are subject to change (e.g., pH, oxygen availability, and nutrient availability), making it difficult to create an environment conducive to the growth and development of the microbial culture. Other drawbacks are the potential for escape of non-native microbial cultures from treatment areas and possible blocking of the aquifer with solids. Also, it may be undesirable to add a primary growth substrate to groundwater, especially if the

water is used as drinking water. Many municipalities will resist adding methane to their water supply, for example.

Some contaminants require a consortium of microbes, for example, polyaromatic hydrocarbons and PCBs. Treatment of these sites requires *bioaugmentation,* or introduction of new microbes. A number of naturally occurring or genetically engineered microbes have been custom-made for specific contaminant removal. These genetically modified bugs have to be introduced into the new environment and induced to grow and eat. The problems in introducing new microbial populations into a given locality have been noted earlier.

Will bioremediation replace the dig-and-burn mentality? Much will depend on public acceptance. Although the concept of bioremediation is generally considered environmentally-friendly, local action groups tend to push for the total removal of the contaminated media. Use of biological clean-up methods will require patience on the part of the public, who tend to just want the stuff gone.

Environmental Applications—Air Emissions

Technology exists to use microorganisms to remove contaminants from air. These technologies have been applied in principle to emissions that range from low concentration of contaminants to flue gases containing sulfur and nitrogen. Methods include use of biofilters or of bioscrubbers, where the gas is sprayed with liquids containing the microbe. Biofilters consist of a support medium, such as sheets of plastic, to which the microbes adhere, and a method to bathe the medium with aqueous nutrients. Some of these methods are surprisingly effective. Treatment of contaminated gases with biofilters has resulted in removal rates of 90 percent, with as low a contaminant concentration as .1–.25 kg organic material per cubic meter (Ratledge and Kristiansen, 2001). However, this technology has been slow to catch on. Problems include the requirement for large floor space, difficulties in pH control, and the fact that the support materials themselves can create odors.

Summary

Cellular enzymes have attracted the interest of industries that produce organic chemicals. This is because in the cell, the enzymes are responsible for controlling the production of cellular organic chemicals, fine-tuning these reactions, and accelerating the reaction rate by factors of hundreds of thousands. Thousands of naturally occurring enzymes have been characterized and are in use in the production of food products, textile dyes, antibiotics, and insecticides. The availability of recombinant DNA techniques provides the opportunity to modify enzymes so that they are less heat-labile, longer-lived, and don't turn off when their product accumulates. It has also provided the opportunity to fuse functions into a single molecule where nature requires several enzymes. Obviously, since this science is new, the potential is just being realized.

To discover new naturally occurring enzymes and to custom-design enzymes is difficult because of the complexity of the cellular soup, and the difficulty of predicting the effect of small changes in the amino acid sequence on the structure and hence the functionality of the protein. This task is made easier by analytical methods based on myriads of data on protein structure and function. Computational methods have grouped proteins throughout the animal and plant kingdoms into superfamilies that have similar structure and similar function. By comparing the structure of a new enzyme to others whose function is known, sometimes the activity of the new enzyme can be extrapolated.

Industrial processes involving cells and cellular enzymes are conducted in bioreactors. Most bioreactors operate in a batch mode where challenges include supplying nutrients and dissolved oxygen, controlling temperature and pH, and eliminating waste. New designs simplify some of these issues with strategies to immobilize the cells or reactants. This allows the reactor to be smaller and simplifies the environmental controls.

Bioremediation techniques have been developed to clean up contaminated soil and water. Naturally occurring microbes will detoxify hazardous waste, although there are strategies to greatly accelerate the rate that they do so, including moisturizing contaminated soil and adding nutrients. Microbes exist naturally or have been developed that digest many organic contaminants and even "eat" metal. The metal-eating bacteria convert metal contaminants to less toxic forms. Groundwater presents a special challenge because of the difficulty of controlling conditions in the aquifer and the fact that the nutrients introduced to support microbial growth are, in themselves, contaminants. In general, bioremediation is less popular with the general public than is the simple removal of the contaminated media from the vicinity. Strategies to use microbes to eliminate toxic air emissions have also been developed.

Quiz

1. Enzymes are:
   (a) cellular proteins that speed up chemical reactions in cells.
   (b) catalysts that can break down other proteins.
   (c) proteins that can serve as switches, moving chemical reactions in different directions.
   (d) naturally engineered to increase or decrease their activity according to cellular conditions.
   (e) all are correct.
2. The use of microbes for chemically contaminated soil:
   (a) requires that the bacteria be genetically modified to eat toxic chemicals or metals.
   (b) requires that the bacteria and the contaminated soil be placed in a bioreactor so that anaerobic conditions can be maintained and the soil can be kept moist.
   (c) requires that the enzymes be used outside of the cell because the hazardous waste kills the bacteria.
   (d) can use naturally occurring microbes.
   (e) all are correct.
3. Bioreactors:
   (a) provide an environment where cells are provided with the optimal amount and types of nutrients, oxygen, and other environmental variables to allow their growth and production of chemicals.
   (b) are devices for killing cells and stripping them of their products.
   (c) are devices to restrain patients while they are infused with biochemicals.
   (d) includes large vats for processing chemical and municipal wastes.
   (e) a and d are correct.

4. The activity of an enzyme can be determined by:
   (a) comparison to similar enzymes whose function you know.
   (b) its high concentration in a cell where the process of interest is ongoing.
   (c) observing the effect of adding the enzyme to the process you are developing.
   (d) all are correct.
5. Enzymes developed through recombinant DNA techniques:
   (a) are inferior to natural products, which have been fine-tuned through the process of evolution.
   (b) can merge functions seen in enzymes among different species.
   (c) can be engineered to withstand high temperatures and pH extremes.
   (d) can be engineered to merge functions of several enzymes into one molecule.
   (e) all are correct.
   (f) b, c, and d are correct.
6. The decision whether to conduct an industrially important process using whole cells or to extract and use the enzymes outside of the cell depends on:
   (a) whether the cells can withstand production conditions.
   (b) whether cofactors must be regenerated.
   (c) whether enzymes from several cell types must be combined.
   (d) whether several enzymes from the same cell are important.
   (e) all are correct.
7. Superfamilies of proteins:
   (a) number in the hundreds of thousands.
   (b) have been identified as occurring between closely related species.
   (c) have similar structures and functions.
   (d) all are correct.
8. Use of biofilters to reduce air emission:
   (a) is not feasible because it is ineffective.
   (b) requires large floor space and may create new odors.
   (c) has been widely adopted.
   (d) b and c are correct.

9. Bioremediation of aquifers:
   (a) is easier than that of soil because the environment is aqueous.
   (b) is easier than that of soil because the conditions are well controlled.
   (c) has met with wide success.
   (d) may require introduction of contaminants into a municipal water supply.
   (e) all are correct.
10. Bioreactors must:
   (a) provide a source of nitrogen.
   (b) supply dissolved oxygen.
   (c) remove waste products.
   (d) control pH.
   (e) All are correct.

References

Crameri, A., S. A. Raillard, E. Bermudez, and W. P. Stemmer. 1998. DNA shuffling of a family of genes from diverse species accelerates directed evolution. *Nature* 391 (6664): 288–291.

Klibanov, Alexander M. 2001. Improving enzymes by using them in organic solvents. *Nature* 409 (6817): 241–246.

Ratledge, C., and B. Kristiansen. 2001. *Basic biotechnology.* 2nd ed. Cambridge, Mass: Cambridge University Press.

Salsh, C. 2001. Enabling the chemistry of life. *Nature* 409 (6817): 226–231.

Schmid, A., J. S. Dordick, B. Hauer, M. Wubbolts, and B. Witholt. 2001. Industrial biocatalysis today and tomorrow. *Nature* 409 (6817): 258–268.

CHAPTER 13

The Future

"Yet, in holding scientific research and discovery in respect, as we should, we must also be alert to the equal and opposite danger that public policy could itself become the captive of a scientific-technological elite."

—President Eisenhower's Farewell Address to the Nation, January 17, 1961

The advances in biotechnology are made possible by the coming of the information age. Without advanced computer technology, we would be severely limited in our analytical techniques and on our ability to synthesize data. We would have progressed at a snail's pace in our efforts to understand the genome. Sequencing the human genome would have taken a generation, if it were possible at all. The union of computer technology and genetic engineering has accelerated the pace of discovery to the point that we as a society are struggling to keep up, legally and ethically.

You might be asking, "If this is so revolutionary, how come I don't see any differences—why am I not affected?" In truth, most of us have not yet seen the effects directly. Food is not especially better (and certainly not cheaper) and medical care has been minimally affected. Employers are not checking our genomes and our friends are not producing designer babies. There are only two possible explanations for this lack of impact: either the biotech revolution is not all that invasive, or we are seeing just the beginning of the impact of biotechnology. My money is on the latter.

The Legal Dilemmas

Being the good citizens that we are, we trust that our government has a handle on technological developments and the regulation thereof. If our metric is the issuance of patents, we can sleep well, because our government is issuing patents to beat the band. Patents have been granted for vaccines, vectors, antigens, antibodies, proteins, gene-encoding proteins, receptors, specific lines of transgenic plants, and the use of gene expression as a diagnostic tool. Tens of thousands of patents have been issued on biological entities and techniques in the last three decades. What is the significance of the explosion in patents applications? Have there really been that many innovations in technology?

One school of thought argues that many of these patents are wrongfully being issued since they are discoveries rather than inventions. According to this line of reasoning, one should no more be able to patent a protein than to patent gravity or subnuclear particles. The fear is that competitive interests are obstructing the usual flow of information among scientists, and patent issues are impeding scientific progress. Researchers may be impeded from truly benefiting from the discoveries of others. Also, researchers are tempted to devise new methods and technologies that may not be necessary solely to avoid paying license fees. The opposing view holds that potential for commercialization, i.e., to make a profit from a discovery, is necessary to drive the industry forward.

The rush for patents is part of the academic as well as the commercial community because the funding for much of the research conducted at universities is through private companies. Without the potential to patent these discoveries, private companies would not have an incentive to fund needed research. Consider the original funding of stem cell research; this was not government-funded. The development of the first stem cell lines resulted in numerous patents to the funding entities, the Wisconsin Alumni Research Foundation (WARF) and the private company, Geron.

Some of the patents are so broad as to fail to establish a clear boundary between where the discovery ends and new research begins. Consider the recent patent issued to Kaufman, S. et al. (U.S. Patent 5,824,514.2), as described by Bialy (1999). The Kaufman patent applies to a method to produce a library of polynucleotide molecules starting with a predetermined property of interest. It seems that such a method should be available to anyone tinkering around in a biochemistry laboratory. But, as a result of this patent, anyone using this method will be subject to a patent fee. As another example, a discoverer of a new application for a known protein may receive a patent for this application. Consider the patent issued to Ferrara et al. (U.S. Patent 6,969,758). It describes a method of modifying cell structure so that the enzyme biliverdin reductase, or "fragments or variants thereof," will be delivered to the site of tissue repair.

The issue of patenting genes is even more peculiar. In 1980, the U.S. Supreme Court awarded to a microbiologist, Ananda Mohan Chakrabarty, a patent for the development of the first oil-eating bacterial strain (even though this strain had already been found to be too fragile for use in the wild). This event sanctioned the commercialization of life forms. Further actions of the courts revealed that genes in people can be patented, and not necessarily by those that might consider themselves the owners of the gene. John Moore was an Alaskan businessman who found that a cell line from his spleen tissue had been patented by the University of California at Los Angeles (UCLA) and licensed to the Sandoz Pharmaceutical Corporation. John's spleen tissue was desirable, apparently, because it produced blood proteins that enable the growth of natural killer cells—a property that was discovered when John was undergoing treatment for cancer. John might have been less irritated if UCLA had mentioned to him that they intended to patent his unique gene. And, in fact, in 1990 the California Supreme Court held UCLA liable for failure to inform Moore of the plans, but upheld the validity of the patent. Since then, many patents have been granted for unique genes found in indigenous human populations and exotic plants and animals. Entire cell populations have been patented, such as the patent held by Biocyte, awarded by the European Patent Office, on human blood cells from the umbilical cords of a newborn child, used for any therapeutic purposes.

In the dark ages, scientists shared their discoveries hoping for fame, if not fortune. Consider the example of the technology used to produce monoclonal antibodies. Ceasar Milstein, working together with George Kohler at the University of Cambridge, developed hybridomas, hybrid cell lines that produce monoclonal Abs. This technology enabled rapid progress in areas related to monoclonal antibodies. In the opinion of Clark (2005), Milstein's decision not to patent a discovery for which he and Kohler received the Nobel Prize contributed to the rapid and wide dissemination of mAb technology. However, universities today have been encouraged to establish their own technology transfer organizations and rules. A discovery like the one made by Milstein and Kohler would most certainly be patented in today's scientific community.

The primary factor that has changed the biomedical research world is the potential to make money—lots of money—from such things as the chance isolation of a small molecule that, for example, helps certain proteins to fold correctly. Researchers are quickly patenting their nucleotide sequences for molecules that they have discovered or fabricated, in hopes they have found the big one. It's human nature.

In contrast to other aspects of biotechnology, the patentability of new plant varieties has well-established precedents. Current patents in the U.S. are based on the U.S. Plant Patent Act of 1930. This act applies to asexually produced plants and allows a claim to the plant variety itself. It excludes others from selling, marketing, offering, delivering, exporting, etc., the crop. A similar certificate under the U.S. Plant Variety Protection Act (PVPA) of 1970 can be obtained for sexually produced or tuber-propagated plant varieties. These patents have a term of 20 years. Plant

patents are not available in Europe, but similar protection is provided through "breeder's rights" (Agris, 1999).

The industry claims that current intellectual property protection laws don't fully control the use of transgenes once they are expressed in seed. Industry sources estimate that in 2000 in the Saskatchewan region of Canada alone, more than 300,000 acres of wheat were planted with unregistered or obsolete GM plant varieties. No harvesting method is capable of containing all of the seeds that are produced. Plants that spring from seeds left in the field are called "volunteers." Controlling volunteers from GM crops would clearly be more difficult than controlling traditional plants because of the resistance to herbicides. At least part of the crises with Bt crops in India is due to the proliferation of illegal, genetically-modified varieties of cotton planted alongside legal varieties (Jayaraman, 2002). So, the control of GM technology is not just a matter of intellectual property protection of the companies developing the GM variety, but also pertains to the safety of this technology and the ability to contain GM genes.

Public Confidence

To brush off public concerns about various aspects of biotechnology as distorted perceptions of risks is a mistake—just ask the nuclear industry. The general public has valid concerns focused on genetic engineering. The discovery of "foreign genes" where they should not be has prompted some extreme reactions, and has resulted in unexpected costs to agricultural entities. The incident with StarLink corn was discussed in Chapter 11. For another example, in 1999, the Swiss Department of Agriculture announced that non-GM corn had been found to be contaminated with foreign genes and burnt thousands of acres of cornfields in Switzerland. Again, in 1999, the European Union (EU) detected in a shipment of honey from Canada the presence of a protein from pollen with an unapproved GM trait. The EU rejected the shipment. Honey shipments to Europe dropped by $4.8 million between 1998 and 2000. Note that after honey is filtered, it contains only about 0.1 percent pollen. Recently, the President of Zimbabwe rejected the offer of food shipments to relieve the devastating famine in his country. Why? Because the food being offered was genetically modified. What people specifically fear is not clear, but they clearly want to know what they are eating.

Loss of public trust can occur without any real indication of risk. For example, a British documentary aired in 1999 reviewed research that supposedly reported stunted growth and repressed lymphocyte function in rats fed GM corn. Although the research was highly suspect, public reaction was intense and there were calls for moratoriums on GM research. A survey of European attitudes conducted in 1999 suggested that

Europeans have become increasingly opposed to genetically modified foods, but they remain supportive of medical and environmental applications (Gaskell et al., 2000).

In the U.S., there appears to be greater acceptance of GM technology than in Europe, but there is a firm commitment to preserve the purity of U.S. agriculture in the realm of organic farming. The U.S. Department of Agriculture recently confirmed that it will not include genetically modified organisms and similar modern biotechnology methods among proposals for nationwide standards for defining organic agriculture practices. This position reflects the public mindset that genetic engineering is unnatural and, by inference, possibly unhealthy. Some companies have invested in consumer resistance to GM products, including Gerber and Archer Daniels Midland, by aggressive advertising of their products as non-GM. Other companies, equally large, have accepted both GM and traditional products.

Public acceptance of biotechnology will probably increase with familiarity. Acceptance of medical advances will come much more easily than will that of agricultural technology. However, as with attitudes toward the nuclear industry, don't expect concerns to evaporate. In fact, even incidences that are small in the mind of the industry can create major, costly setbacks.

The Issue of Labeling

One index of the level of public concern is the rigor of labeling requirements. Labeling requirements in the United States are issued by the Food and Drug Administration. Currently, requirements are to label food as genetically modified only if the product has a significantly different nutritional property or contains an allergen that consumers might not expect. However, the FDA has proposed voluntary guidelines for labeling. Most of the volunteers to date have been companies that have *not* used GM products and are anxious to so inform their customers.

The EU, by contrast, plans to establish a tracking system that would mandate labeling for both foods containing transgenic ingredients and foods that were derived from transgenic material, such as oils. Such labeling requirements are widely supported by the European public. There is even debate over requiring labeling of animals that were fed GM food.

The issue of labeling is far from simple. If labeling is required, there would be some minimum level of GM constituency established to trigger the labeling requirements—below regulatory concern if you will. Based on experience with other food additives, the reporting level will be set well below any risk-based concerns. Companies would be obligated to establish testing regimens to prove that their products met the criteria. This would be costly, as would potential lawsuits and loss of business, if a given commodity was found to contain above the regulated level of

GM product. However, to date, 22 countries have announced plans to institute some form of mandatory labeling. Polls show that most U.S. consumers support labeling even if it means an increase in price.

Genetic Pollution

Concerns over the movement of any genetically modified trait into the biosphere have been raised in both scientific and public arenas. Numerous studies confirm that unexpected movement of modified genetic traits is not only a credible concern (Dale et al., 2002; Petra et al., 2003), but, in fact, has already occurred. Migration of specific genetic traits could result in herbicide-resistant plants or insecticide-resistant insects.

In 1999, the first triple-resistant canola was discovered in Alberta. The strain is resistant to Roundup, Liberty, and Pursuit. Industry had not developed the triple-resistant crop; the genetic material had been exchanged between traditional crops and GM crops to give rise to this triple-resistant plant. Canola is an open-pollinating crop. That means that genders exchange genetic material in pollinating seeds and also means that transfer of genetic material is fairly easy. While it might be acceptable or even desirable for canola to be so hardy, the development of such traits in the related weed population, wild mustard, would be disastrous. Weeds resistant to the same herbicides as the crops would move into crop areas. Roundup-resistance has moved into some wild grass populations. Roundup-tolerance has been demonstrated in annual rye grass in Australia and described in a recently published example in the United States (Dale et al., 2002). Note that natural evolutionary pressures also work to increase herbicide-resistance among weedy plants.

Another facet of the same concern involves the loss of genetic diversity. If genetic diversity of the global agricultural species is diminished, one can argue that the entire globe might be affected by a new or more newly aggressive pathogen. The spread of corn blight across the U.S. in 1970 is an example of the potential consequences of the lack of genetic diversity. Because all species of corn were closely related, even in the 1970's, the corn crop sustained a huge loss from the blight. Of course, humans had dramatically diminished genetic diversity in agriculture long before genetic engineering was available, though selective breeding practices. Nonetheless, new technology makes this process much more rapid and potentially more globally distributed.

A highly publicized incident involved the discovery of transgenic corn in remote southern Mexico (Quist and Chapela, 2001). Southern Mexico is the center of diverse species of maize. The study by Quist and Chapela has been widely criticized, especially their conclusion that the new DNA could integrate into the genome at multiple sites. If random insertion can occur, this little gene could wreak havoc by turning on genes not normally active and deactivating necessary genes. Follow-up studies have confirmed Quist and Chapela's findings of the migrated transgene, although the more serious claim of random insertion has not been validated.

In response to public concerns, the Mexican government has established a moratorium on the growing of transgenic corn. However, Mexico still imports a large quantity of GM corn intended as food from the U.S. The Mexican farmer can purchase this corn and plant it rather than eat it. In addition, corn falls from trucks going to market and can sprout along the roadside.

How big of a problem is gene migration? Reported problems are minimal to date. A significant concern is the fact that much of the control of GM technology depends on the cooperation of farmers. Dale et al. (2002) report that in the year 2000, almost 30 percent of farmers failed to comply with the refuge protocols designed to prevent or delay the emergence of insects resistant to Bt toxins. The problem is aggravated by the implementation of the technology across different cultures. And the fact that the controls are costly, such as the requirement of a commitment of valuable acreage to "refuge" plots. Controls engineered into the plants will be more effective, such as terminator technology.

There are other technologies on the horizon that could cause different concerns. For example, technology has been developed to interfere with lignin biosynthesis in trees, decreasing the crop rigidity and the cost of processing wood to produce products like paper. However, lack of rigidity is clearly an undesirable quality in a long-lived species. Migration of traits such as this into the wild population might seriously damage forest and would certainly create public resistance to GM technology in general.

Bioethics

Historically, ethics and codes of behavior evolved over stretches of time as civilizations grew and changed. In today's world, ethical challenges are thrown at us so fast that there is no time for societal adjustment and development of a consensus about what is right and what is wrong. Instead, ethics are issued as decrees, usually from some branch of government. Consider the example of drug use. It is fair to say that our civilization is in violent disagreement about the ethics of recreational drugs. While many parents (but unfortunately far from all) and the school and the legal systems discourage drug use of any kind, the pop culture glorifies the use of drugs through very well-funded marketing.

Biotechnology is likely to be the granddaddy of all ethics challenges because we are altering the substance of life. While it might be very comfortable to take the

position that the manipulation of genes is wrong, this position is not going to be useful because the genie is already out of the bottle. It's here and it's growing.

The President's Council on Bioethics is chartered to advise the president on ethical issues related to advances in biomedical science and technology. (Edmund D. Pellegrino, M.D., Council Chairman). This council maintains a web site (http://www.bioethics.gov/) where comprehensive information is compiled regarding the bioethical issues in front of us. The dialogue runs the gamut, from futuristic applications to the ethical use of Ritalin. Among the concerns raised is the potential of biotechnology to extend the human life span and thereby distort human demographics. Also raised is the potential use of "baby engineering" for purposes ranging from sex selection to projected performance on an SAT exam. Will muscle stem cells be used to create superathletes? It works in mice. The engagement in dialogue is certainly healthy, but the options of a government entity for dealing with bioethics are truthfully limited. For example, fetal screening on the basis of sex (mostly based on selective abortion) is a current reality, and the gender ratios are already significantly skewed in some areas of the world. The problems this may create are obvious; what to do about it is less obvious.

One pressing issue is the length of time that biotechnological techniques are taking to move from the laboratory to the clinic. The fantastic potential offered by new options is well known to the public. Many spinal cord injury victims believe that the U.S. government stands between them and a potential cure. Such patients are easy targets for clinics outside of the country, some of which are of dubious credibility. Some clinics operated overseas not only offer credible medical care but may provide critical data on the effectiveness of these treatments. Sheridan (2005) discusses the practice of Amit Patel, the director of the Center for Cardiac Cell Therapy at the University of Pittsburgh Medical Center. He injects stem cells from his patients' bone marrow into their hearts to treat cardiac disease, not at the facility in Pittsburgh but at the Bangkok Heart Hospital in Thailand. The data gleaned from these studies could substantively improve future therapies and might accelerate the approval process in this country, where conservatism rules the day. The FDA has recently authorized a company named StemCells, Inc. to inject fetal neural stem cells into the brains of children suffering from Batten disease, a devastating neurodegenerative disorder. Geron plans to file an application soon to perform clinical trials using cells derived from embryonic stem cells in the treatment of spinal cord injury. Although authorization of clinical trials is an important first step, it is often years before such therapies are approved for general use.

A significant problem for stem cell therapy, compared to bioengineered molecules, is that the cost structure has no mechanism to compensate developers for the costly process of developing these treatments and bringing them into the clinic. We may reasonably apply money from the public sector to subsidize stem cell therapies. The NIH recently pledged $6.5 million to help move stem cell therapies into the clinic.

There is a danger in progressing too fast. Sheridan (2005) quotes David Beck, head of the Cornell Institute for Medical Research in Camden, New Jersey. Dr Beck notes that overly high expectations could breed cynicism that would set the field back. For example, Viacell Gamada recently suspended a Phase 1-2 trial of umbilical cord blood therapy for blood cancers because two of eight enrolled participants developed severe graft-versus-host disease. The immunological complications were unexpected because the cord blood is reputedly much less antigenic than adult blood. Arguably, the clinical trial was premature. More time spent to produce better science before launching into clinical trials might actually save time and money.

The dilemma over how rapidly to push radical, bioengineered technology into the clinic is typical of our bioethical concerns. We have technology that can relieve human suffering and should be pushed forward. However, if wrongly or carelessly used, unprecedented consequences may be experienced on a global scale. Even if the technology is designed with the best intentions, such as to improve geriatric medicine, the outcome may be questionable: consumption of huge societal resources in the care of the elderly. As a bioengineer, your best bet is to encourage full and honest dialogue with the public, thereby ensuring, at least, that value-based decisions are made in the public sector, not in a company boardroom.

Developing Nations

Major agricultural corporations typically emphasize the potential for their new technology to "feed the world." Advertisements note the burgeoning world population and imply an opportunity to somehow make a dent in the miserable state of much of this population, even as it grows. Certainly, hardier plants, more nutritious plants, and greater crop yield could provide relief to hard-pressed areas. However, the immediate impact is not always positive. Remember the events with the Malaysian palm oil industry discussed in Chapter 11? This industry is the mainstay of the Malaysian economy and is based on the extraction of lauric acid from palm nuts grown in Malaysia. This industry has been seriously threatened by the technology to induce the production of lauric acid by rapeseed plants. Lauric acid, the primary product from palm oil, can be produced much more cheaply by the rapeseed plant grown throughout the world. Consider the scenario of simply increasing the yield of a primary crop in an agricultural economy. The immediate impact may be to throw a sizeable fraction of the population out of work, if the crop can be produced and/or harvested with fewer workers.

The developing world is ambivalent about GM technology to say the least. Some parts of the developing world have avidly embraced this technology, while other areas are suspicious. In 2000, Sri Lanka banned all GM foods. The majority of GM crops

outside of the industrialized West are in Argentina and South Africa, planted entirely by larger farmers for export (Fletcher, 2001). What a scenario! Brazil has marketed the fact that their soy is not GM, but Brazilian farmers are smuggling seeds in from nearby Argentina. The Brazilian government has responded by jack-boot tactics—raiding storage sheds, burning fields, etc. Meanwhile, soy prices are at an all-time low.

The primary contributor to world hunger is probably not global productivity but rather food distribution (Fletcher, 2001). However, biotechnology could be certainly helpful on a local scale by allowing crops to grow in areas that previously were unproductive. The economic incentive to develop such technology and to push it into areas where it is needed is less than the incentive for technology to be sold to large corporations for big-money crops. Furthermore, depressed areas are less well positioned to regulate this technology and to set up monitoring systems.

The Bt crises in India has yielded some important lessons in dealing with technology development in developing nations. Local climate and culture must be carefully considered before GM technology is exported. As early as 2002, private groups were claiming that the Monsanto Bt corn was not as productive in India as had been expected, even as the Indian government was claiming great successes with this technology (Jayaraman, 2002). By 2005, a publication of a study by a government institute in India revealed that the Bt cotton planted in India is not as efficient in killing bollworms there as it is in the U.S. or China because it is not designed for India's longer ripening season (Jayaraman, 2002). This sequence of events lead to angry charges that Monsanto was falsely promoting the use of hybrids in India to induce farmers to buy fresh seeds every year even though native varieties actually perform better. The damage to Monsanto's public image in India will be very difficult to repair.

The case of golden rice will be a good test of how successfully new technology can be integrated with local crops and can be provided to those who most need it. Golden rice has been genetically modified to produce beta carotene and potentially to alleviate vitamin A deficiencies in developing nations. The inventors of golden rice have given Syngenta the exclusive license to charge for the commercial use of the golden rice technology. However, the terms of the agreement with Syngenta should guarantee that the original motives of the inventors in developing this technology are not compromised. According to the agreement, Syngenta will support the humanitarian use of golden rice and all subsequent applications of the technology. For implementation, a "golden rice humanitarian board" has been established to determine priorities and to ensure compliance with regulations governing GM crops. The board has established partnerships with the Philippines, India, China, Vietnam, Indonesia, and Africa. The partnering institutions will set up a framework that best suits the local region and will conduct studies on local factors that might affect this crop and its usefulness. Importantly, the partnering institutions will also transfer the trait into locally adapted lines and ensure availability to the most needy of their population. The wise implementation of the golden rice technology may contribute to international goodwill by showing that Western technology can and will be used to improve the general living conditions in developing nations.

Worst Case Scenarios

We have been through some pretty grim scenes: killer grass, indestructible weeds and insects, lack of diversity among major crops followed by a major disaster, control of the world's food supply by a handful of megacompanies, designer babies, use of stem cells to create superathletes, and denial of opportunities based on individual DNA sequence, among others. But consider the potential of this technology to relieve human suffering! Empathize with the parents of children with genetic diseases, or of children blinded by Vitamin A deficiency. Consider the potential for stem cells to treat ischemic heart disease and congestive heart failure, the major causes of morbidity and mortality in the U.S. Certainly, we have done some things right and many things wrong. And certainly this technology could go awry. It's up to you, bioengineer, to turn this technology to the greater good.

Summary

The ability to decipher the genetic code and to genetically modify plants and animals has the potential to change the world. However, in its infancy, this technology has created some legal and ethical dilemmas. The issuance of patents to life forms, to molecules, and to DNA sequences has been controversial because these entities were discoveries, not inventions. However, protection of intellectual property by patents is argued to be essential to provide incentives for development and commercialization of biotechnology. And, to judge by the number of patents that have been issued, incentive certainly exists. Ethically, we struggle with fears about the potential of biotechnology. Will our DNA really be the subject of scrutiny by potential employers? Will we screen our embryos based on DNA characteristics? Will we really live into our second century and crowd out the young? Our ability to deal with these questions will depend to a large part on how successful we are as a society in making value-based decisions in the full light of public scrutiny.

The availability of GM crops has created some unprecedented environmental issues. There is widespread fear of genetic pollution of natural crops or of related wild species. The fear is that herbicide resistance will spread, that insects will develop resistance to insecticide, and that somehow eating an unnatural gene will be harmful. There have been some widely publicized and costly incidences of

foreign genes migrated into "pure" crops or into native species, giving credence to these concerns. Of further concern is the fact that GM technology is being exported into regions where the technology is unlikely to be closely regulated.

The developing world has been ambivalent about GM crops. Several countries have banned GM crops altogether, although there appears to be a very healthy black market. The incident with Bt cotton in India illustrates the importance of considering local culture and climate before exporting GM crops. The failure of the Bt cotton to thrive in the longer growing seasons of India has created widespread suspicion of Western technology in general and of Monsanto in particular. The more cautious implementation of golden rice technology may be a better model for global dispersion of a GM trait. Golden rice is being dispersed through partnerships with local institutions. The local partners are commissioned to introduce the trait through appropriate local varieties and to ensure the technology is available to those who most need it.

Quiz

1. Patents:
   (a) in the area of biotechnology, must be based on something the applicator has invented.
   (b) have been granted on individual antigens.
   (c) have been granted on DNA sequences.
   (d) have been granted on new uses for a known enzyme.
   (e) b, c, d are correct.
2. Genetic pollution:
   (a) is the only way that a weedy plant may develop resistance to a herbicide.
   (b) has been shown to not be a hazard with GM crops.
   (c) has been discovered to have occurred in wild Mexican maize.
   (d) is a healthy and economical way for crops to develop multiple herbicide resistance.
   (e) is unlikely in developing countries because agriculture is so diverse.

3. Golden rice:
   (a) is being implemented worldwide in American varieties of rice.
   (b) will be marketed solely as a big-money crop through large agricultural corporations.
   (c) is being implemented through insertion of the trait in local rice varieties.
   (d) has migrated to wild rice varieties.
   (e) a and b are correct.
4. The creation of superathletes:
   (a) is theoretically possible by injection of muscle stem cells.
   (b) has been shown experimentally in mice.
   (c) has been endorsed by the President's Commission on Bioethics.
   (d) all are correct.
   (e) a and b are correct.
5. GM crops in developing nations:
   (a) have enjoyed widespread popularity.
   (b) have alleviated hunger in many parts of the world.
   (c) have been banned in several developing nations.
   (d) have been economically beneficial wherever the technology has been implemented.
   (e) a, b, and d are correct.
6. Moving stem cell therapy to clinical applications:
   (a) has occurred more rapidly in some other countries compared to the U.S.
   (b) has recently received some funding from the public sector.
   (c) is beginning in the United States, as evidenced by recent consideration or approval of several clinical trials.
   (d) could damage the ultimate use of this technology if done too rapidly.
   (e) all are correct.

7. Labeling of GM products:
   (a) is more rigorous in the U.S. than elsewhere in the world.
   (b) will save money.
   (c) reflects the degree of concern of the public regarding GM technology.
   (d) all are correct.
8. Loss of genetic diversity:
   (a) began in agriculture with the introduction of GM technology.
   (b) could result in greater vulnerability of major crops and wildlife to the development of a new pathogen.
   (c) could decrease the ability of wildlife and major crops to withstand climate change.
   (d) all are correct.
   (e) b and c are correct.
9. Public perception of risk:
   (a) may not be based on scientific reasoning and therefore should be ignored.
   (b) may result in huge costs to the industry over a seemingly minor event.
   (c) is a problem for biotechnology in the Western world but not in developing countries.
   (d) is easily addressed by effective advertising.
10. Biotechnology:
    (a) has been embraced by the whole world as a positive development.
    (b) has been better received in medical developments than in agricultural developments.
    (c) has not engendered government regulations except in the United States.
    (d) has been promoted primarily in third-world countries.

References

Agris, C. H. 1999. Intellectual property protection for plants. *Nature Biotechnology* 17 (2): 197–198.

Bialy, H. 1999. Intellectual property and the parsing of protein space. *Nature Biotechnology* 17 (1): 2–3.

Clark, M. 2005. Empowering the inventor—the case of monoclonal antibodies. *Nature Biotechnology* 23 (9): 1047–1049.

Dale, P. J., B. Clarke, and E. M. Fontes. 2002. Potential for the environmental impact of transgenic crops. *Nature Biotechnology* 20 (6): 567–574.

Fletcher, L. 2001. GM crops are no panacea for poverty. *Nature Biotechnology* 19 (9): 797–798.

Gaskell, G., N. Allum, M. Bauer, J. Durant, A. Allansdottir, H. Bonfadelli, D. Boy, et al. 2000. Biotechnology and the European public. *Nature Biotechnology* 18 (9): 935–938.

Quist, D., and I. H. Chapela. 2001. Transgenic DNA introgressed into traditional maize landraces in Oaxaca, Mexico. *Nature* 414 (6863): 541–543.

Jayaraman, K. S. 2002. Poor crop management plagues Bt cotton experiment in India. *Nature Biotechnology* 20 (11): 1069.

Petra, Meier and Wilfried Wackernagel. 2003. Monitoring the spread of recombinant DNA from field plots with transgenic sugar beet plants by PCR and natural transformation of *Pseudomonas stutzeri. Transgenic Research* 12 (3): 293–304.

Sheridan, C. 2005. Cord blood cell therapy trial suspended. *Nature Biotechnology* 23 (12): 1455–1456.

Final Exam

1. The cell membrane forms a micelle because:
   (a) the cell contains proteins that fold the membrane correctly.
   (b) hydrostatic pressure outside of the cell causes the membrane to bend.
   (c) the sugar heads of the molecules forming the membrane aggregate, and the lipid tails also aggregate, resulting in a two-layer structure.
   (d) the membrane is pushed out by osmotic pressures inside the cell.
2. Cellular energy:
   (a) is produced primarily in the mitochondria of eukaryotic cells.
   (b) is twice as efficient in the presence of oxygen.
   (c) requires oxygen.
   (d) is based on the chemical process whereby carbon is altered from a reduced to a more oxidized form.
   (e) all are correct.
   (f) a and d are correct.
   (g) a, c, and d are correct.

3. Nucleic acids:
   (a) form a double-stranded helix.
   (b) include RNA, DNA, and other molecules such as ATP.
   (c) contain bases.
   (d) contain amino acids.
   (e) a, b, and d are correct.
   (f) b and c are correct.
4. Lipids:
   (a) are the primary source of cellular energy.
   (b) are characterized by loving water (hydrophilic).
   (c) have a charged surface.
   (d) are the basis for high-energy storage molecules.
   (e) all are correct.
5. Proteins:
   (a) carry the genetic information code for the cell.
   (b) are composed of hundreds of different kinds of amino acids.
   (c) store energy.
   (d) compose structural elements.
   (e) all are correct.
6. Hydrogen bonds:
   (a) are a type of covalent bonds.
   (b) are stronger than typical ionic bonds.
   (c) are due to the formation of polarity in covalently bonded molecules.
   (d) a and c are correct.

7. Potential uses of stem cells that have at least been demonstrated in laboratory animals include:
   (a) development of super-animals through injection of muscle stem cells.
   (b) improvement of spinal cord injury.
   (c) improvement of cardiac diseases through injection of bone marrow stem cells.
   (d) doubling of life span.
   (e) improvement in skin graft techniques.
   (f) all except a and d are correct.
   (g) all except d are correct.
   (h) all are correct.
8. Energy flow within the cell:
   (a) oxidizes carbon to produce heat.
   (b) transfers energy from oxidation of carbon to high-energy phosphate bonds.
   (c) transfers the energy from oxidation of hydrogen to high-energy phosphate bonds.
   (d) requires as input a six-carbon sugar.
   (e) all are correct.
   (f) all except a are incorrect.
   (g) all except a and c are incorrect.
9. Energy storage within the organism:
   (a) is in the form of fat.
   (b) is in the form of the carbohydrates starch or glycogen.
   (c) is in the form of high-energy phosphate bonds.
   (d) all are correct.
   (e) all except c are incorrect.
10. Embryonic stem cells:
    (a) are derived from cultivating a fertilized egg.
    (b) are derived from aborted fetuses.
    (c) are derived by dissecting a 16-cell zygote.
    (d) are developed from a small population of cells in the center of a 150-cell blastocyte.
    (e) any of these techniques would work.

11. Consequence of misfolded proteins include:
    (a) accumulation of amyloids.
    (b) inability to secrete product.
    (c) change in function.
    (d) inability to bind as a receptor.
    (e) all are correct.
    (f) all except a are correct.
12. Concerns regarding use of embryonic stem cells include that they might:
    (a) differentiate into the wrong type of cell if injected into a human organ.
    (b) develop into a baby.
    (c) give rise to cancer.
    (d) induce a graft versus host reaction.
    (e) carry contaminants from the cultivation layer of mouse cells.
    (f) all are correct.
    (g) all except b and e are correct.
    (h) all except b are correct.
13. A DNA chip:
    (a) is a microelectronic circuit programmed to analyze a DNA sequence.
    (b) is an array of DNA segments from a known cell in a known state.
    (c) is an array designed to capture labeled DNA segments complementary to the segments on the chips.
    (d) can be used to determine which genes have been "turned on" in a given cell.
    (e) all are correct.
    (f) all except a and c are correct.
    (g) all except a are correct.
14. Mutations:
    (a) are transmitted to future generations if the cell survives.
    (b) often result in cells that are not viable or are destroyed by natural killer cells.
    (c) only damage the organism if they occur in germ cells.
    (d) all are correct.

15. The human genome is unusual among animals in the:
    (a) number of genes.
    (b) unique DNA codes used to transcribe for proteins.
    (c) amount of DNA consisting of repeating segments.
    (d) vulnerability to mutations.
    (e) all are correct.
16. Gene therapy is an option if:
    (a) the genetic basis for the disease is well understood.
    (b) the disease is due to a missing enzyme.
    (c) the disease is due to a misfolded protein.
    (d) a strategy is available to apply the gene to the appropriate cells.
    (e) all are correct.
    (f) a and c are correct.
17. Prions are:
    (a) nucleotide sequences that allow proper excretion of proteins.
    (b) molecules that enable proper folding of proteins.
    (c) misfolded proteins.
    (d) catalytic molecules that "edit" mRNA.
18. The concerns with the movement of transgenes outside of the GM crop include:
    (a) loss of genetic diversity within related agricultural and wild species.
    (b) consumption of an unidentified product by the consumer.
    (c) development of herbicide resistance by weedy plants.
    (d) development of insecticide resistance by insects feeding on numerous Bt-producing species.
    (e) all are correct.
    (f) all except b are correct.

19. The molecules in the cell's membrane:
    (a) have a polar head and a non-polar tail.
    (b) form a double-layered circle.
    (c) form a micelle.
    (d) exclude polar molecules from the cell interior.
    (e) exclude charged molecules from the cell interior.
    (f) b, c, and d are correct.
    (g) b, c, and e are correct.
20. Which of the following is true of polymers?
    (a) Biomolecules consist of long strings of like units.
    (b) The function of a polymer may be changed by exchanging a few units.
    (c) Units are built by molecules, like or unlike, hooked together like LEGOs®.
    (d) Proteins are not considered polymers.
    (e) All are true.
    (f) b and c are true.
21. Ribozymes are:
    (a) proteins that have catalytic activity (can attack other molecules).
    (b) sites for protein synthesis.
    (c) small nucleotides.
    (d) molecules that stop DNA duplication read-out if they are incorporated into a new DNA code.
22. Proteins are unique among molecules because they:
    (a) form structural units.
    (b) can edit mRNA.
    (c) contain peptide bonds.
    (d) contain nitrogen.
    (e) store energy.

23. mRNA:
    (a) produces transfer RNA which produces protein.
    (b) ensures the proper folding of proteins.
    (c) "edits" proteins to create the proper amino acid sequence.
    (d) can be induced to transcribe a protein sequence in the presence of reverse transcriptase.
    (e) all are correct.
24. You would cultivate a cell in media containing ampicillin after incubating with plasmids containing recombinant DNA because:
    (a) the host cell is resistant to ampicillin and contaminating cells will be killed.
    (b) the ampicillin-resistant gene is on the plasmid; surviving cells have incorporated the plasmid.
    (c) the ampicillin-resistant gene is on the recombinant DNA; surviving cells have incorporated the DNA.
    (d) the ampicillin-resistant gene is a reporter gene.
25. Eukaryotic cells have the following advantage over prokaryotic cells for recombinant DNA technology:
    (a) They are easy to cultivate.
    (b) They have a simple genome.
    (c) They can perform the post-translational operations on proteins.
    (d) They can act as vectors.
    (e) They can be harvested in batch quantities.
    (f) c and e are correct.
26. Polymerase chain reaction is a method of:
    (a) building protein polymers.
    (b) sequencing DNA.
    (c) producing large quantities of a given DNA fragment from a small sample.
    (d) creating DNA from mRNA populations.

27. You would cultivate a cell in media containing X-gal after incubating with plasmids containing recombinant DNA because:
    (a) the host cell is resistant to X-gal and contaminating cells will be killed.
    (b) the beta-galactosidase gene is on the plasmid; colored cells have incorporated the plasmid.
    (c) the beta-galactosidase gene is on the recombinant DNA; colored cells have incorporated the DNA.
    (d) the beta-galactosidase gene is a reporter gene.
28. Options for vectors include:
    (a) viruses.
    (b) plasmids.
    (c) chloroplasts.
    (d) mitochondria.
    (e) a and b are correct.
    (f) all are correct.
29. One way to ensure that the protein that you want is transcribed (read out) from the recombinant DNA is to:
    (a) insert the gene that you want transcribed in-between a known start sequence and a known stop sequence on the plasmid.
    (b) incorporate a reporter gene.
    (c) insert the gene onto a plasmid that contains an ampicillin-resistant gene.
    (d) incubate in the presence of reverse transcriptase.
30. The difference between RNA polymerase and DNA polymerase is that:
    (a) RNA polymerase binds at specific sites on the DNA.
    (b) RNA polymerase binding requires the DNA to be in a specific configuration.
    (c) DNA polymerase initiates the duplication of DNA.
    (d) DNA polymerase requires the DNA to be in a specific configuration.

31. The discovery of restriction enzymes was important in the development of recombinant DNA technology because restriction enzymes:
    (a) act as vectors and insert foreign DNA into the host.
    (b) cut DNA into small fragments that may be inserted into the host DNA.
    (c) edit mRNA to produce the final protein.
    (d) cut DNA at specific nucleotides so that complementary DNA may be added at a known point.
32. A stem cell is:
    (a) a cell that can develop into any other cell in the organism.
    (b) a cell that can divide and grow and develop into the organism.
    (c) a cell derived from the human embryo.
    (d) a relatively undifferentiated cell that can develop into at least several different cell types.
33. DNA fingerprinting is based on:
    (a) matching DNA sequences.
    (b) karyotypes.
    (c) the banding patterns on the chromosomes.
    (d) how may chromosomes are found.
34. How would you find a specific molecule within a cell?
    (a) Use monoclonal antibodies against that molecule labeled with a fluorescent or radioactive label.
    (b) Use polyclonal antibodies labeled with a fluorescent or radioactive label.
    (c) Incubate the cell with radioactive tritium.
    (d) Screen for messenger RNA coding for that molecule.
35. Sickle cell anemia victims:
    (a) can benefit from bone marrow stem cell transplants because the defect is in the RBCs.
    (b) suffer from a recessive genetic disease.
    (c) could potentially benefit from gene therapy.
    (d) derive primarily from areas with high incidence of malaria.
    (e) all are correct.
    (f) all except a are correct.

36. Consider an incomplete dominance model using chrysanthemums—red (R), white (r), and pink (Rr). Which of the following is not true?
    (a) RR × Rr produces no white flowers.
    (b) Rr × rr produces no red flowers.
    (c) Rr × Rr produces no pink flowers.
    (d) RR × rr produces no white flowers.
37. Which of the following is true about color blindness:
    (a) A color-blind mother will produce all color-blind daughters.
    (b) •A color-blind mother will produce all color-bind sons.
    (c) A color-blind father will produce all color-blind sons.
    (d) A color-blind father will produce all color-blind daughters.
38. What is a germ cell?
    (a) A fertilized egg
    (b) Another name for a stem cell
    (c) Sperm or unfertilized egg
    (d) A mutated cell
39. Chicken eggs are a promising source of human proteins because:
    (a) the chicken is somewhat protected from proteins secreted into their eggs.
    (b) generation times of chickens are short.
    (c) chickens are highly productive.
    (d) eggs are a concentrated source of protein.
    (e) all are correct.
40. Impediments to distributing GM technology to developing nations include:
    (a) incompatibility with local climates.
    (b) suspicion on the part of local people.
    (c) lack of economic incentive for companies.
    (d) inability or unwillingness to comply with use controls for GM crops.
    (e) all are correct.

41. GM crops have:
    (a) been used primarily to benefit needy populations.
    (b) greatly improved the economies of developing nations.
    (c) saved money for farmers.
    (d) resulted in lower food prices.
    (e) all are correct.
42. The primary control to prevent development of resistance to Bt as a result of widespread use of GM crops that secret Bt is:
    (a) spraying Bt on weedy plants.
    (b) ensuring that the Bt is of sufficient quantity to kill all the insects.
    (c) planting sizeable acreage (refuge) with non-Bt plants.
    (d) using additional insecticides to supplement the Bt.
43. Antisense technology could be used to treat viral-induced diseases such as HIV because the antisense molecule:
    (a) scrambles the viral DNA codes, preventing its readout.
    (b) is complementary to the single stranded RNA of the virus.
    (c) prevents the virus from entering the cell.
    (d) causes misfolding of the protein produced by the virus.
44. Why would you want to produce human proteins in plants?
    (a) Using animal proteins limits supply.
    (b) Using animal proteins might cause an immunological reaction.
    (c) Plants can produce human proteins.
    (d) Using plants is cheaper than harvesting proteins from non-genetically engineered animals.
    (e) all are correct.
45. Non-food crops are preferred for producing therapeutic human proteins because:
    (a) consumers fear the introduction of therapeutic human proteins in their food.
    (b) human proteins might be toxic to wildlife.
    (c) production of the food would compete with production of the therapeutic protein.
    (d) patients will not be willing to receive their medication as part of their food.

46. Gene sequencing might be able to detect early cancers because pre-cancer cells:
    (a) have a lot of mutations.
    (b) express oncogenes, some of which have been characterized.
    (c) are infected with viruses.
    (d) have their own blood supply.
47. The complement system:
    (a) amplifies the immune response.
    (b) is initiated by antibodies.
    (c) is nonspecific.
    (d) depends on the Fc portion of the antibody.
    (e) all are correct.
48. The quaternary structure of proteins is important because:
    (a) it determines protein configuration and therefore the ability to act as an enzyme.
    (b) misshapen proteins are marked for disposal.
    (c) protein secretion depends on proper configuration.
    (d) misshapen proteins can form harmful products in the cell.
    (e) all are correct.
    (f) a, b, and c are correct.
49. Current rDNA products:
    (a) have not yet appeared in the clinic.
    (b) consist solely of mAbs to treat cancer and autoimmune disease.
    (c) are costly.
    (d) b and c are correct.
50. Characteristics of cancer cells include:
    (a) drug resistance.
    (b) long telomers.
    (c) uncontrolled growth.
    (d) unique surface antigens.
    (e) all are correct.
    (f) a, b, and c are correct.

51. Reasons to grow GM crops include:
    (a) insect resistance.
    (b) pesticide resistance.
    (c) drought resistance.
    (d) increased nutritional value.
    (e) all are correct.
    (f) a, b, and c are correct.
52. Strategies to use mABs to combat autoimmune diseases might include antibodies against:
    (a) CD4 receptors.
    (b) interleukin 12.
    (c) MHC II.
    (d) the autoimmune antibodies.
    (e) all are correct.
    (f) all except c are correct.
53. In vitro fertilization (IVF) and DNA sequencing will currently allow you to select for offspring according to:
    (a) intelligence.
    (b) height.
    (c) eye color.
    (d) certain genetic diseases.
54. Why does DNA fingerprinting sometimes not convince juries?
    (a) People don't trust science.
    (b) More than one person can exhibit the same DNA fingerprint.
    (c) The process is vulnerable to contamination.
    (d) DNA fingerprinting requires a large sample.
    (e) All are correct.
    (f) b and c are correct.

55. Potential treatments for HIV include:
    (a) preventing read-out of the viral genes by antisense molecules.
    (b) blocking the CD4 receptor.
    (c) using gene therapy to replace the defective HIV gene.
    (d) blocking HIV receptors on lymphocytes.
56. Aptamers:
    (a) are produced in response to the presence of a specific protein.
    (b) shield hydrophobic surfaces.
    (c) shield hydrophilic surfaces.
    (d) edit mRNA.
57. Nucleotides consist of all of the following except:
    (a) sugar.
    (b) phosphate.
    (c) organic bases.
    (d) lipids.
    (e) carbohydrates.
58. Which of the following is true about inheritance of height?
    (a) Inheritance follows Mendelian genetics.
    (b) Tall is dominant.
    (c) Short is dominant.
    (d) Height is affected by hormonal factors, such as human growth factor.
    (e) a, b, and d are correct.
59. Which of the following is true about applying therapeutic proteins?
    (a) They break down in the GI tract.
    (b) Treatment is expensive.
    (c) It is difficult to deliver to the site where they are needed.
    (d) Intravenous application is required unless they are encapsulated.
    (e) All are correct.

60. DNA polymerase:
    (a) initiates cell division.
    (b) produces DNA from mRNA.
    (c) assembles the correct nucleotides.
    (d) responds to signals that initiate cell division.
    (e) all are correct.
    (f) c and d are correct.
61. Problems with producing proteins in transgenic animals include:
    (a) the animals are slow to reach maturity.
    (b) products may be toxic to the animals.
    (c) the animals must be killed to obtain the protein.
    (d) the animals cannot fold the human proteins correctly.
    (e) a and b are correct.
    (f) a, b, and d are correct.
62. Cell division is stopped if:
    (a) chromosomes don't line up properly.
    (b) mutations are detected.
    (c) proteins are misfolded.
    (d) telomers are not aligned.
63. A reason that a protein would not be secreted might be:
    (a) hydrophilic surfaces are exposed.
    (b) lack of secretory sequences on the protein.
    (c) missing chaperones.
    (d) lack of appropriate receptors.
    (e) all are correct.
    (f) a and b are correct.
    (g) a, b, and c are correct.

64. Gene therapy can currently cure:
    (a) spinal cord injury.
    (b) diabetes.
    (c) SCIDS.
    (d) MS.
    (e) all are correct.
65. Oxidation of carbon:
    (a) requires ATP.
    (b) requires oxygen.
    (c) releases energy.
    (d) can release hydrogen.
    (e) all are correct.
    (f) b, c, and d are correct.
    (g) c and d are correct.
66. The Human Genome Project:
    (a) is scheduled to be complete in 2010.
    (b) is revealing significant genetic differences between races.
    (c) will reveal the function of all human genes.
    (d) is showing that all mammals have approximately the same number of genes.
    (e) all are correct.
67. Gamma globulins:
    (a) consist of antibodies.
    (b) are applied as non-specific enhancement of the immune response.
    (c) are a type of lipid.
    (d) are polyclonal antibodies.
    (e) all are correct.

68. Concerns about the potential for biotechnology to decrease plant diversity include:
    (a) New pathogens will destroy all related species.
    (b) Monotony among food products.
    (c) A few megacompanies will control the seed supply.
    (d) Aesthetics.
    (e) All are correct.
    (f) a, c, and d are correct.
69. T-cells:
    (a) mature in the bone marrow.
    (b) mature in the thymus.
    (c) produce antibodies.
    (d) include dendritic cells.
    (e) b, c, and d are correct.
    (f) b and d are correct.
70. B-cells:
    (a) mature in the bone marrow.
    (b) mature in the thymus.
    (c) produce antibodies.
    (d) include dendritic cells.
    (e) a, c, and d are correct.
    (f) a and c are correct.
71. Ubiquitin:
    (a) enables secretion.
    (b) initiates DNA synthesis.
    (c) edits mRNA.
    (d) marks misfolded proteins for destruction.

72. The access of mABs to cells:
    (a) depends on the antibodies binding to specific receptors on the surface.
    (b) is improved with radioactive labeling of the antibody.
    (c) is improved if only the Fab fragment is used.
    (d) all are correct.
73. You would add oxygen to a bioreactor for an anaerobic reaction to:
    (a) kill the reaction.
    (b) stimulate product production.
    (c) stimulate cell growth.
    (d) stop cell growth.
    (e) a and c are correct.
    (f) b and d are correct.
74. President Bush's announcement regarding federal funding of research on stem cells in 2001:
    (a) greatly decreased the federal funding available for embryonic stem cell research.
    (b) greatly increased the federal funding available for embryonic stem cell research.
    (c) prohibited adult stem cell research.
    (d) prohibited human cloning.
    (e) a and d are correct.
75. A biotech company would be interested in the genome of indigenous people because:
    (a) genetic differences could justify eugenics.
    (b) slight differences in the genome could explain differences in disease incidence.
    (c) genetic reasons for differences in cancer incidence might be discovered.
    (d) biotech companies are notoriously interested in anthropology.
    (e) b and c are correct.

76. A biotech company would be interested in the proteomic of exotic plants because:
    (a) it might want to grow and market the exotic plant.
    (b) it might discover a protein with therapeutic value.
    (c) it might want to insert a transgene.
    (d) it want to ensure flora diversity.
77. Options for producing a protein include:
    (a) harvesting the mRNA from a cell producing the protein.
    (b) using a DNA synthesizer to create the proper DNA sequence from the amino acid sequence.
    (c) chemically synthesizing the protein.
    (d) all are correct.
78. The John Moore case was significant legally because it determined that:
    (a) we all are legal owners of our own genome.
    (b) genes cannot be patented.
    (c) companies and medical institutions can patent the genes of individuals.
    (d) companies must show value added before genes can be patented.
79. The Chakrabarty case was significant legally because it determined that:
    (a) life forms cannot be patented.
    (b) life forms can be patented.
    (c) companies or individuals must show value added before life forms can be patented.
    (d) life forms can be patented only if they are genetically modified.
80. Which of the following types of molecules may be used for structural elements?
    (a) Lipids and carbohydrates
    (b) Proteins, carbohydrates, and nucleotides
    (c) Lipids, proteins, and carbohydrates
    (d) Proteins and carbohydrates

81. In cell division (mitosis):
    - (a) chromosomes divide once and the cell divides twice.
    - (b) the chromatin divides before chromosomes are visible.
    - (c) there is a period where heterologous chromosomes are aligned together and exchange genetic material.
    - (d) mutations often result.
    - (e) all are correct.
82. Cells regulate the entry and exit of material by:
    - (a) excluding large nonpolar molecules.
    - (b) excluding large polar molecules.
    - (c) requiring that many compounds enter by first attaching to specific receptors.
    - (d) preventing diffusion of small molecules.
    - (e) requiring that molecules to be excreted have specific secretary sequences.
    - (f) all except a are correct.
    - (g) all except b are correct.
    - (h) b, c, and e are correct.
83. The way that biotechnology may provide help for Alzheimer's sufferers is:
    - (a) gene therapy for the defective gene.
    - (b) enzyme therapy.
    - (c) vitamins.
    - (d) aptamers to improve folding.
84. Catalysts:
    - (a) edit mRNA.
    - (b) accelerate speed of reactions.
    - (c) cut apart proteins.
    - (d) decrease the energy required for a reaction.
    - (e) all are correct.
    - (f) all except a are correct.

85. If you consider the making of a protein analogous to cooking, which molecule is the cook?
    (a) mRNA
    (b) DNA
    (c) tRNA
    (d) RNA polymerase
    (e) Ribosomes
86. Which is not true of GM fish?
    (a) GM fish are currently being marketed.
    (b) GM fish are of concern because of the history of hatchery fish escaping into the wild.
    (c) GM fish are hardier and survive better in the wild than native fish.
    (d) GM fish grow bigger and are more resistant to disease than native fish.
    (e) All are correct.
87. The Bt protein is of concern because:
    (a) it is not natural and therefore harmful to human health.
    (b) widespread use may lead to Bt-resistant insects.
    (c) it kills butterflies.
    (d) it kills birds.
    (e) all are correct.
    (f) b and c are correct.
88. The controversy regarding ice minus was that:
    (a) it is not natural and therefore harmful to human health.
    (b) widespread use may lead to ice minus resistance.
    (c) it may affect the climate.
    (d) it may decrease flora diversity.
89. Without an endoplasmic reticulum, a cell would be:
    (a) unable to conduct oxidative metabolism.
    (b) less efficient in protein production and excretion.
    (c) unable to destroy proteins.
    (d) unable to divide.

90. Thalassemia is an example of a disease where:
    (a) the normal gene of the gene pair can compensate for a defect.
    (b) both genes must be normal.
    (c) the defect is in a cell receptor.
    (d) the defect is a missing enzyme.
91. StarLink corn was pulled off the market at a huge cost to the producing company because:
    (a) it contained products known to be toxic to humans.
    (b) it contained pesticides.
    (c) it was genetically modified and therefore unsuitable for human consumption.
    (d) it had not been approved for human consumption.
92. Dominant genetic diseases:
    (a) are transmitted to at least 50 percent of the offspring.
    (b) are predominantly transmitted to sons.
    (c) are predominantly transmitted to daughters.
    (d) are transmitted to 100 percent of the offspring.
93. In the analogy of cooking and proteins, which molecule obtains the recipe for the cook?
    (a) mRNA
    (b) RNA polymerase
    (c) DNA polymerase
    (d) tRNA
    (e) Ribosomes
94. A problem with inserting genes into living DNA is that it might:
    (a) interrupt an existing gene, converting it into nonsense.
    (b) not be expressed.
    (c) create an oncogene.
    (d) create a chimera.
    (e) all are correct.
    (f) a, b, and c are correct.
    (g) a and c are correct.

95. Which of the following is true about the transcription of a human gene?
    (a) RNA polymerase attaches at a known organic base start sequence.
    (b) The mRNA produces the same protein from a read-out of the same sequence of organic bases every time.
    (c) Each protein has its own gene.
    (d) Every sequence of three organic bases translates into a given amino acid.
    (e) All are correct.
    (f) a and d are correct.
    (g) a, c, and d are correct.
96. Adult stem cells:
    (a) are found in the umbilical cord.
    (b) are capable of developing into any of the over 200 cell types of the body.
    (c) have been used to grow new human hearts.
    (d) all are correct.
    (e) b and c are correct.
97. In germ cell division (miosis):
    (a) chromosomes divide once and the cell divides twice.
    (b) there is a period where heterologous chromosomes are aligned together and exchange genetic material.
    (c) mutations may result.
    (d) all are correct.
    (f) a, b, and c are correct.
98. In a bioreactor, why would you want your cells to adhere to spheres?
    (a) Easier distribution of nutrients.
    (b) Easier oxygenation.
    (c) Easier to eliminate waste.
    (d) More surface area for cell growth.
    (e) All are correct.

99. Which of the following is true about biotechnology?
    (a) Fledgling biotechnology companies have usually thrived.
    (b) It is easier to go to clinical trials with rDNA products in Europe compared to the United States.
    (c) Many rDNA drugs are in the approval process.
    (d) Numerous rDNA drugs have been cancelled late in the approval process.
    (e) All are correct.
    (f) a, b, and c are correct.
    (g) b, c, and d are correct.
100. Dendritic cells are removed from victims of kidney cancer, exposed to antigens from the cancer cells, and reinjected into the patient because the dendritic cells:
    (a) need to be taught that the cancer antigens are not self.
    (b) will present the cancer antigens to T-cells and initiate an immune response.
    (c) will seek out and kill the cancer cells.
    (d) will initiate the complement cascade.
101. Somatic nuclear transfer:
    (a) is a method for introducing a transgene into a somatic cell.
    (b) is a method for harvesting product from a somatic cell.
    (c) is a method for producing clones.
    (d) is a method of detecting pre-cancerous cells.
    (e) none are correct.

Answers to Quiz and Exam Questions

Chapter 1: Biomolecules and Energy

1. d
2. f
3. f
4. g
5. b
6. e
7. e
8. a
9. e
10. g

Chapter 2: Cell Structures and Cell Division

1. b
2. d
3. e
4. b
5. f
6. e
7. a
8. e
9. e
10. f

Chapter 3: Information Methods of a Cell

1. c	2. c	3. a	4. e	5. e
6. c	7. e	8. d	9. a	10. e

Chapter 4: Genetics

1. a	2. b	3. d	4. b	5. c
6. d	7. e	8. f	9. b	10. b

Chapter 5: Immunology

1. e	2. b	3. e	4. d	5. c
6. b	7. c	8. d	9. f	10. d

Chapter 6: Immunotherapy and Other Bioengineering Applications

1. d	2. b	3. c	4. f	5. f
6. e	7. d	8. a	9. e	10. d

Chapter 7: Recombinant Techniques and Deciphering DNA

1. b	2. e	3. e	4. e	5. c
6. f	7. e	8. e	9. f	10. c

Chapter 8: Proteomics

1. e	2. a	3. e	4. d	5. e
6. f	7. c	8. d	9. d	10. d

Chapter 9: Stem Cells

1. f	2. b	3. f	4. g	5. b
6. g	7. d	8. e	9. d	10. e

Chapter 10: Medical Applications

1. a	2. e	3. c	4. e	5. c
6. d	7. e	8. f	9. b	10. f

Chapter 11: Agricultural Applications

1. e	2. c	3. e	4. f	5. b
6. e	7. e	8. c	9. e	10. b

Chapter 12: Industrial and Environmental Applications

1. e	2. d	3. e	4. d	5. f
6. e	7. c	8. b	9. d	10. e

Chapter 13: The Future

1. e	2. c	3. c	4. e	5. c
6. e	7. c	8. e	9. b	10. b

Final Exam

1. c	2. f	3. f	4. d	5. d
6. c	7. g	8. f	9. d	10. d
11. e	12. f	13. g	14. b	15. c
16. f	17. c	18. e	19. f	20. f
21. b	22. c	23. d	24. b	25. f
26. c	27. d	28. f	29. a	30. c
31. d	32. d	33. a	34. a	35. e
36. c	37. b	38. c	39. e	40. e
41. c	42. c	43. b	44. e	45. a
46. b	47. e	48. e	49. c	50. e
51. e	52. e	53. d	54. f	55. a
56. b	57. d	58. d	59. e	60. f
61. e	62. a	63. b	64. c	65. g
66. d	67. e	68. b	69. f	70. f
71. d	72. c	73. e	74. b	75. e
76. b	77. d	78. c	79. b	80. d
81. b	82. h	83. d	84. b	85. e
86. e	87. f	88. c	89. b	90. b
91. d	92. a	93. a	94. f	95. f
96. a	97. d	98. e	99. g	100. b
101. c				

INDEX

F

G

N

Q

R

S

T

U

V

W

X

Y

Z